Ex Libris

MATHEMATICAL PREPARATION FOR GENERAL CHEMISTRY

WILLIAM L. MASTERTON
PROFESSOR OF CHEMISTRY,
UNIVERSITY OF CONNECTICUT, STORRS, CONNECTICUT

EMIL J. SLOWINSKI
CHAIRMAN, DEPARTMENT OF CHEMISTRY
MACALESTER COLLEGE, ST. PAUL, MINNESOTA

W. B. SAUNDERS COMPANY
Philadelphia   London   Toronto

 Saunders Golden Series

W. B. Saunders Company: West Washington Square
                           Philadelphia, Pa. 19105

                           12 Dyott Street
                           London, WC1A   1DB

                           833 Oxford Street
                           Toronto, Ontario M8Z 5T9, Canada

Mathematical Preparation for General Chemistry   SBN 0-7216-6174-2 .

Print No.:   9   8   7   6   5

PREFACE

The past decade has been one of fundamental changes in the teaching of mathematics at the secondary school level. Greater emphasis has been placed upon the logic and underlying principles of the discipline. We see this reflected in the capabilities of our students in general chemistry. Compared to their predecessors of a few years ago, they show a greater aptitude for the quantitative reasoning required in problem solving.

Perhaps inevitably, this trend has led to a decreased emphasis on the techniques of mathematics. We are pleased that our students can explain what a logarithm is, but it disturbs us that so few of them can use a table of logarithms with any facility. While many of our students are familiar with the laws of probability, they are seldom able to apply the rules governing the use of significant figures.

Students are capable of filling these gaps in their mathematical background on their own. This text is designed as a guide for a program of self-study in the simple mathematical techniques essential to success in general chemistry. It features a large number of examples and problems, for which answers are provided (Appendix 3), usually with suggested methods of solution. The application to *chemistry* is stressed throughout; nearly all of the problems are phrased in the language of chemistry.

The first seven chapters cover the mathematical background common to virtually every course in general chemistry. They discuss exponents (Chapter 1), logarithms (Chapter 2) and significant figures (Chapter 4). The use of the slide rule and the conversion factor approach to solving problems are presented in Chapters 3 and 5, respectively. Chapter 6 (Algebraic Equations) stresses approximation methods of solving the types of equations that arise in connection with weak electrolyte equilibria. A discussion of simple techniques of graphing is included in Chapter 7 (Functional Relationships).

It is not intended that these chapters be "covered" in lecture, or even "assigned" in the usual sense. Rather, we would expect the student who is having difficulty with the calculations of stoichiometry to familiarize himself with the conversion factor approach by working the problems in Chapter 5. An instructor who finds that his students have never heard of significant figures can refer them to Chapter 4.

Much of the material in Chapters 8 through 11 will not be required in the typical general chemistry course. Instructors who wish to use the principles of trigonometry (Chapter 8) or calculus (Chapters 9 and 10) may refer their students to selected topics in these areas. The statistical treatment of experimental error (Chapter 11) should be of interest to students in quantitative analysis, whether or not this topic is included in the general chemistry course. Chapter 12 is intended to give the reader an understanding of how computers work and an appreciation of their versatility. If he becomes interested enough to visit the computer center early in his college career, we will have accomplished our purpose.

To close on a personal note, we would like to dedicate this book to the memory of two men who made elementary mathematics come alive for one of us (W.L.M.). They are Ellis McKeen of Kennett High School, Conway, N.H., and Marvin Solt of the University of New Hampshire.

WILLIAM L. MASTERTON

EMIL J. SLOWINSKI

CONTENTS

CHAPTER 1

EXPONENTIAL NUMBERS

In chemistry, we frequently deal with very large or very small numbers. In one gram of the element carbon there are:

$$50,150,000,000,000,000,000,000$$

atoms of carbon. At the opposite extreme, the mass of a single carbon atom is:

$$0.00000000000000000000001994 \text{ grams}$$

Numbers such as these are not only difficult to write; they are very awkward to work with. Imagine how tedious it would be to find the mass of 2150 carbon atoms by carrying out the operation:

$$0.00000000000000000000001994 \text{ grams}$$
$$\times\ 2150$$

One way to simplify operations of this type is to use what we call **exponential notation.** In exponential notation, numbers such as those written above are expressed as a number between one and ten (**coefficient**) times an integral power of ten (**exponential**). Examples of exponential numbers include:

$$\text{Coefficient} \longleftarrow\quad 1 \times 10^4; \quad 2.23 \times 10^3; \quad 5.6 \times 10^{-4}$$

To understand precisely what these exponential numbers mean, it may be helpful to refer to Table 1.1.

1

Table 1.1   Exponentials

$$
\begin{aligned}
10^{6} &= (10)(10)(10)(10)(10)(10) &= 1{,}000{,}000 \\
10^{5} &= (10)(10)(10)(10)(10) &= 100{,}000 \\
10^{4} &= (10)(10)(10)(10) &= 10{,}000 \\
10^{3} &= (10)(10)(10) &= 1000 \\
10^{2} &= (10)(10) &= 100 \\
10^{1} &= (10) &= 10 \\
10^{0} &= &= 1 \\
10^{-1} &= (0.1) &= 0.1 \\
10^{-2} &= (0.1)(0.1) &= 0.01 \\
10^{-3} &= (0.1)(0.1)(0.1) &= 0.001 \\
10^{-4} &= (0.1)(0.1)(0.1)(0.1) &= 0.0001
\end{aligned}
$$

For the three numbers cited previously, we have:

$$1 \times 10^{4} = 1(10{,}000) = 10{,}000$$

$$2.23 \times 10^{3} = 2.23(1000) = 2230$$

$$5.6 \times 10^{-4} = 5.6(0.0001) = 0.00056$$

1.1   WRITING NUMBERS IN EXPONENTIAL NOTATION

In order to make use of exponential notation, we must be able to write any number, large or small, as an exponential number. To understand how this is done, it may be helpful to start with two relatively simple cases which can be worked directly from the entries in Table 1.1.

Suppose we wish to express the number 5196 in exponential notation. We realize that this number can be written as:

$$5.196 \times 1000$$

Referring to Table 1.1, we note that $1000 = 10^{3}$. Therefore:

$$5196 = 5.196 \times 1000 = 5.196 \times 10^{3}$$

As another illustration, consider the number 0.0028. To express this number in exponential notation, we first write it as:

$$2.8 \times 0.001$$

Since $0.001 = 10^{-3}$, we have:

$$0.0028 = 2.8 \times 0.001 = 2.8 \times 10^{-3}$$

The method which we used to work these two examples is not particularly useful with extremely large or extremely small numbers, for which tables of exponentials are seldom available. We can, however, deduce from these examples a more general approach to the problem. Notice that when we expressed 5196 in exponential notation, we arrived at an exponent of 3; this is the *number of places which the decimal point must be moved (to the left) to give the coefficient, 5.196.* Again, in expressing 0.0028 in exponential notation, the exponent, -3, is the *number of places which the decimal point must be moved (to the right) to give the coefficient, 2.8.* In general:

To express a number in exponential notation, write it in the form:

$$C \times 10^n$$

where C is a number between 1 and 10 (e.g., 1, 2.62, 5.8) and n is a positive or negative integer (e.g., 1, -1, -3). To find n, count the number of places that the decimal point must be moved to give the coefficient, C. If the decimal point must be moved to the left, n is a positive integer; if it must be moved to the right, n is a negative integer.

Example 1.1 Express the two numbers given at the beginning of this chapter (the number of atoms in one gram of carbon and the mass in grams of one carbon atom) in exponential notation.

Solution For the number:

$$50,150,000,000,000,000,000,000$$

the coefficient is 5.015. To obtain this coefficient, the decimal point must be moved 22 places (count them!) to the *left*. It follows that the exponential number is:

$$5.015 \times 10^{22}$$

Similarly, the coefficient of the number:

$$0.000000000000000000000001994$$

is 1.994. The decimal point must be moved 23 places to the *right* to obtain the coefficient. Therefore, we obtain:

$$1.994 \times 10^{-23}$$

EXERCISES

Express the following numbers in exponential notation:

1. 1000
2. one billion
3. 0.000001
4. 16,220
5. 212.6
6. 0.189
7. 6.18
8. 0.00000007846

1.2 MULTIPLICATION AND DIVISION

One of the principal advantages of exponential notation is that it greatly simplifies the processes of multiplication and division. In applying these processes to exponential numbers, we make use of the fact that *to multiply, we add exponents:*

$$10^1 \times 10^2 = 10^{1+2} = 10^3$$
$$10^6 \times 10^{-4} = 10^{6+(-4)} = 10^2$$

To divide, we subtract exponents:

$$10^3/10^2 = 10^{3-2} = 10^1$$
$$10^{-3}/10^6 = 10^{-3-6} = 10^{-9}$$

Using these principles, we arrive at the following rules for multiplying or dividing exponential numbers:

To multiply one exponential number by another, first multiply the coefficients together in the usual manner. Then add exponents.

To divide one exponential number by another, divide coefficients in the usual manner and subtract exponents.

Example 1.2 Carry out the indicated operations:
   a. $(5.00 \times 10^4) \times (1.60 \times 10^2)$
   b. $(6.01 \times 10^{-3})/(5.23 \times 10^6)$

Solution

  a. For convenience, we first separate the coefficients from the exponential terms:

$$(5.00 \times 1.60) \times (10^4 \times 10^2)$$

Multiplying coefficients and adding exponents, we obtain: 8.00×10^6. (Here, and throughout this chapter, we use the rules discussed in Chapter 4 to express answers to the correct number of significant figures.)

  b. $(6.01 \times 10^{-3})/(5.23 \times 10^6) = \dfrac{6.01}{5.23} \times \dfrac{10^{-3}}{10^6} = 1.15 \times 10^{-9}$

We frequently find that when exponential numbers are multiplied or divided, our answer is not in standard exponential notation. Consider, for example:

$$(5.0 \times 10^4) \times (6.0 \times 10^3)$$

Carrying out this multiplication in the usual manner, we obtain:

$$(5.0 \times 6.0) \times (10^4 \times 10^3) = 30 \times 10^7$$

Again:

$$(3.60 \times 10^2)/(4.92 \times 10^4) = \dfrac{3.60}{4.92} \times \dfrac{10^2}{10^4} = 0.732 \times 10^{-2}$$

The two numbers just obtained, 30×10^7 and 0.732×10^{-2}, are not expressed in standard exponential form, since the coefficients are *not* numbers between 1 and 10. To express these numbers in exponential notation, we follow the procedure described in Example 1.3.

Example 1.3 Express the numbers 30×10^7 and 0.732×10^{-2} in standard exponential notation (i.e., as numbers between 1 and 10, times 10 to the proper power).

Solution In the first case, we write:

$$30 \times 10^7 = (3.0 \times 10^1) \times 10^7$$

Adding exponents, we obtain: 3.0×10^8

What we did here was to divide the coefficient by 10 so as to obtain a number, 3.0, which lies between 1 and 10. To compensate for this, we multiplied the exponential term, 10^7, by 10 to get 10^8.

Following an analogous procedure in the second case:

$$0.732 \times 10^{-2} = (7.32 \times 10^{-1}) \times 10^{-2} = 7.32 \times 10^{-3}$$

Here, we multiplied the coefficient by 10 and divided the exponential by 10, leaving the value of the number unchanged.

EXERCISES

Carry out the indicated operations.

1. $(6.20 \times 10^4) \times (1.50 \times 10^8)$
2. $(4.3 \times 10^{-3}) \times (9.0 \times 10^4)$
3. $(3.62 \times 10^4) \times (2.91 \times 10^{-7})$
4. $(8.16 \times 10^{-4}) \times (4.78 \times 10^{19})$
5. $(1.39 \times 10^7)/(1.10 \times 10^4)$
6. $(3.48 \times 10^3)/(6.72 \times 10^5)$
7. $(7.2 \times 10^{-3})/(3.6 \times 10^4)$
8. $(2.60 \times 10^4)/(7.70 \times 10^{-12})$
9. $\dfrac{(6.10 \times 10^4) \times (3.18 \times 10^{-4})}{(8.08 \times 10^7) \times (1.62 \times 10^{11})}$

1.3 RAISING TO POWERS AND EXTRACTING ROOTS

To raise an exponential number to a power, we take advantage of the fact that:

$$(10^a)^b = 10^{ab}$$

To illustrate this rule, consider:

$$(10^2)^3 = 10^2 \times 10^2 \times 10^2 = 10^6 = 10^{(2 \times 3)}$$
$$(10^{-2})^4 = 10^{-2} \times 10^{-2} \times 10^{-2} \times 10^{-2} = 10^{-8} = 10^{(-2 \times 4)}$$

To raise an exponential number to a power, we treat the coefficient in the usual manner and raise the exponential term according to the rule just given:

$$(2.0 \times 10^{-3})^2 = (2.0)^2 \times (10^{-3})^2 = 4.0 \times 10^{-6}$$

The same principle can be used to extract a root (square root, cube root, etc.). Here, we are dealing with a fractional power ($\frac{1}{2}$, $\frac{1}{3}$, etc.), but the principle is the same.

$$(10^6)^{1/2} = 10^{(6 \times 1/2)} = 10^3$$

$$(4.0 \times 10^6)^{1/2} = (4.0)^{1/2} \times (10^6)^{1/2} = 2.0 \times 10^3$$

As before, we operate on the coefficient and exponential term separately.

Extracting roots poses a special problem when the resultant exponent would not be a whole number. Consider, for example:

$$(4.0 \times 10^5)^{1/2}$$

If we follow the procedure outlined above, we obtain:

$$(4.0)^{1/2} \times (10^5)^{1/2} = 2.0 \times 10^{5/2}$$

Our answer is not in standard exponential form; indeed, $10^{5/2}$ is an extremely awkward expression to work with because it cannot readily be translated into an ordinary number.

In cases of this type, we first transform the number we are to operate on into a form such that extracting the root of the exponential term will give a whole number. In the above example, we write:

$$(4.0 \times 10^5)^{1/2} = (40 \times 10^4)^{1/2}$$

multiplying the coefficient by 10 and dividing the exponential by 10. Now, on extracting the square root, we obtain:

$$(40 \times 10^4)^{1/2} = 40^{1/2} \times (10^4)^{1/2} = 40^{1/2} \times 10^2 = 6.3 \times 10^2$$

(The square root of 40 may be found from tables, by the use of logarithms [Chapter 2], or with the aid of a slide rule [Chapter 3].)

From this example, we deduce that, before attempting to extract the square root of an exponential number, we must first make sure that the power of 10 is divisible by 2 to give an integer. If the exponent is an odd number (e.g., -1, 3, 5), we convert it to an even number (-2, 2, 4) by dividing by 10, simultaneously multiplying the coefficient by 10. With cube roots, we ensure that the exponent is divisible by 3.

$$(2.7 \times 10^7)^{1/3} = (27 \times 10^6)^{1/3} = 27^{1/3} \times (10^6)^{1/3}$$

$$= 27^{1/3} \times 10^2 = 3.0 \times 10^2$$

Example 1.4 Perform the indicated operations.
a. $(6.2 \times 10^{-4})^2$
b. $(3.0 \times 10^6)^{1/2}$
c. $(2.81 \times 10^{-5})^{1/2}$

Solution

a. $(6.2 \times 10^{-4})^2 = (6.2)^2 \times 10^{-8} = 38 \times 10^{-8} = 3.8 \times 10^{-7}$
(Note that in order to obtain the answer in standard form, we divided the coefficient by 10 and multiplied the exponential term by 10.)

b. $(3.0 \times 10^6)^{1/2} = (3.0)^{1/2} \times 10^3 = 1.7 \times 10^3$
(The square root of 3.0 is approximately 1.7.)

c. Here, we must first convert the expression to get an exponent which gives an integer when divided by 2. One way to do this is to divide the exponential term by 10 and multiply the coefficient by 10.

$$(2.81 \times 10^{-5})^{1/2} = (28.1 \times 10^{-6})^{1/2} = (28.1)^{1/2} \times (10^{-6})^{1/2}$$

$$= 5.30 \times 10^{-3}$$

EXERCISES

1. $(2.16 \times 10^{-3})^2 = ?$
2. $(4.9 \times 10^4)^3 = ?$
3. $(6.0 \times 10^{-21})^2 = ?$
4. $(9.0 \times 10^6)^{1/2} = ?$
5. $(8.4 \times 10^5)^{1/2} = ?$

6. $(6.2 \times 10^{-2})^{1/2} = ?$
7. $(1.62 \times 10^{-7})^{1/2} = ?$
8. $(2.14 \times 10^{10})^{1/3} = ?$
9. $(6.0 \times 10^4)^2 \times (3.0 \times 10^{-7})^{1/2} = ?$
10. $(3.0 \times 10^7)^{1/5} = ?$

1.4　ADDITION AND SUBTRACTION

Occasionally, we find it necessary to add or subtract two exponential numbers. These processes are extremely simple if both exponents are the same. To add:

$$2.02 \times 10^7 + 3.16 \times 10^7$$

we factor to obtain: $(2.02 + 3.16) \times 10^7 = 5.18 \times 10^7$

Again,

$$6.1 \times 10^{-5} - 3.0 \times 10^{-5} = (6.1 - 3.0) \times 10^{-5} = 3.1 \times 10^{-5}$$

If the exponents are not the same, the numbers must be operated upon to make the exponents the same. This procedure is illustrated in Example 1.5.

Example 1.5   Carry out the indicated operations.
   a. $6.04 \times 10^3 + 2.6 \times 10^2$
   b. $9.82 \times 10^{-4} - 8.2 \times 10^{-5}$

Solution

a. We cannot perform the addition directly, any more than we can add six oranges to two apples. In order to carry out a meaningful addition, we must make the exponents of the two terms the same. One way to do this is to operate on the second number, expressing it as a coefficient times 10^3.

$$2.6 \times 10^2 = 0.26 \times 10^3$$

Therefore:

$$6.04 \times 10^3 + 2.6 \times 10^2 = 6.04 \times 10^3 + 0.26 \times 10^3 = 6.30 \times 10^3$$

b. Proceeding as in part (a):

$$9.82 \times 10^{-4} - 8.2 \times 10^{-5} = 9.82 \times 10^{-4} - 0.82 \times 10^{-4}$$
$$= 9.00 \times 10^{-4}$$

Alternatively, we could have operated on the first number rather than the second:

$$9.82 \times 10^{-4} = 98.2 \times 10^{-5}$$

Hence:

$$9.82 \times 10^{-4} - 8.2 \times 10^{-5} = 98.2 \times 10^{-5} - 8.2 \times 10^{-5}$$
$$= 90.0 \times 10^{-5} = 9.00 \times 10^{-4}$$

EXERCISES

Carry out the indicated additions and subtractions.
1. $3.02 \times 10^4 + 1.69 \times 10^4$
2. $4.18 \times 10^{-2} + 1.29 \times 10^{-2}$
3. $6.10 \times 10^4 + 1.0 \times 10^3$
4. $5.9 \times 10^{-5} + 1.86 \times 10^{-4}$
5. $8.17 \times 10^5 - 1.20 \times 10^4$
6. $6.49 \times 10^{-10} - 1.23 \times 10^{-11}$
7. $9.68 \times 10^4 + 7.01 \times 10^2$
8. $6.02 \times 10^{23} - 1.0 \times 10^2$

PROBLEMS

These problems are designed to illustrate the use of exponential numbers in general chemistry. Don't panic if you are unfamiliar with some of the vocabulary that is used. A knowledge of exponents combined with a little common sense will enable you to work all of the problems. You may even learn some chemistry in the process!

The answers to the problems are given in Appendix 3, in standard exponential notation.

1.1   A helium atom has the following properties:
>   mass = 0.0000000000000000000000000665 g
>   radius = 0.0000000093 cm
>   average velocity at 25°C = 136,000 cm/sec
>   Express these quantities in exponential notation.

1.2   Using the data in Problem 1.1, calculate:
>   a. The mass, in grams, of Avogadro's number (6.02×10^{23}) of helium atoms.
>   b. The average kinetic energy of a helium atom at 25°C. K. E. $= mv^2/2$, where m = mass, v = velocity.

1.3   When one gram of carbon burns to form carbon dioxide, 7.86×10^3 calories of heat is evolved. Calculate the amount of heat evolved when one mole (12.0 g) of carbon burns. This quantity is called the molar heat of combustion.

1.4   One mole (6.02×10^{23} molecules) of water weighs 18.0 g. Calculate the mass in grams of a water molecule.

1.5   The smallest sample of magnesium that can be weighed on an ordinary analytical balance has a mass of about 1.0×10^{-4} g. How many atoms are there in this sample, given that a magnesium atom weighs 4.0×10^{-23} g?

1.6   A sample of water vapor occupying 2.06×10^4 cc is in equilibrium with 1.2×10^3 cc of liquid water. What is the total volume of the system?

1.7   Dalton's Law tells us that the total pressure of a mixture of hydrogen and helium is the sum of the partial pressures of the two gases. If the total pressure of the mixture is 1.224×10^3 torr and the partial pressure of helium is 9.80×10^2 torr, what is the partial pressure of hydrogen?

1.8   The solubility of helium in water at 25°C is 3.8×10^{-5} moles He/cc water. One mole of helium weighs 4.0 g and has a volume of 2.5×10^4 cc at 25°C. Express the solubility of helium in:

        a. g He/cc water

        b. cc He/cc water

1.9   According to the Bohr theory of the hydrogen atom, the orbital radius is given by:

$$r = \frac{n^2 h^2}{4\pi^2 m e^2}$$

where r is the radius in cm, $h = 6.6 \times 10^{-27}$, $e = 4.8 \times 10^{-10}$, $m = 9.1 \times 10^{-28}$, $\pi = 3.14$, and n is the so-called quantum number which can take on any positive, integral value. Calculate r when $n = 1$; $n = 2$; $n = 3$.

1.10   Coulomb's Law tells us that the force of attraction between two ions in water solution is given by:

$$f = \frac{Z_1 Z_2 e^2}{D(r_1 + r_2)^2}$$

where f is the force (in dynes), Z_1 and Z_2 are the charges of the ions, $e = 4.8 \times 10^{-10}$, $D = 78$, and r_1 and r_2 are the radii of the ions. What is the force between a Na^+ and a Cl^- ion, which have radii of 0.95×10^{-8} cm and 1.81×10^{-8} cm respectively?

1.11   The metal calcium crystallizes in a cubic structure in which the face diagonal, d, of the cube is 7.88×10^{-8} cm. Calculate the length, l, of a side of the cube, using the relation $2 l^2 = d^2$.

1.12   The half life, $t_{1/2}$, of a radioactive isotope is the time required for one half of a sample to decay. It is related to the rate constant for radioactive decay, k, by the equation:

$$t_{1/2} = 0.693/k$$

For the common isotope of radium, $k = 4.33 \times 10^{-4}$/yr. What is the half life of this isotope in:

        a. Years

        b. Days

        c. Minutes

1.13 In a water solution in equilibrium with lead chloride, the concentrations of Pb^{2+} and Cl^- are related by the expression:

$$(\text{conc. } Pb^{2+}) \times (\text{conc. } Cl^-)^2 = 1.7 \times 10^{-5}$$

Calculate the concentration of Pb^{2+} in solutions in which the concentration of Cl^- is:

      a. 1.0×10^{-1}
      b. 2.0×10^{-2}
      c. $2(\text{conc. } Pb^{2+})$

1.14 When ammonia is added to silver chloride, $AgCl$, some of the Ag^+ ions are converted to the $Ag(NH_3)_2^+$ complex. At equilibrium, the following equation holds:

$$\frac{(\text{conc. } Ag^+) \times (\text{conc. } NH_3)^2}{(\text{conc. } Ag(NH_3)_2^+)} = 4 \times 10^{-8}$$

What must the concentration of NH_3 be to make:

      a. conc. $Ag(NH_3)_2^+$ = conc. Ag^+
      b. conc. $Ag(NH_3)_2^+$ = 10^2 (conc. Ag^+)

1.15 In a solution formed by dissolving hypochlorous acid, $HClO$, in water, the following relation exists between the concentrations of H^+ and $HClO$:

$$(\text{conc. } H^+)^2 = (3.2 \times 10^{-8})(\text{conc. } HClO)$$

Complete the following table:

(conc. HClO)	(conc. H^+)	(conc. H^+)/(conc. HClO)
1.0	_____	_____
0.10	_____	_____
0.010	_____	_____

The ratio in the last column represents the fraction of $HClO$ which ionizes.

CHAPTER 2

LOGARITHMS

It was pointed out in Chapter 1 that the processes of multiplication, division, raising to a power, and extracting a root are simplified by expressing the numbers involved in exponential notation. However, even when this is done, a considerable amount of arithmetic may still be necessary. For example, suppose we wish to multiply 6.02×10^{23} by 1.99×10^{-24}. Combining the exponential terms, we arrive at the expression:

$$(6.02 \times 1.99) \times 10^{-1}$$

but we still have to carry out the tedious operation of multiplying 6.02 by 1.99. We could achieve a further simplification if we could express the numbers 6.02 and 1.99 as powers of ten. The time consuming operation of multiplication could then be completely replaced by the more rapid process of addition.

Clearly, what we need is a table which will enable us to express any number as a power of ten. Such a table has been known to mathematicians for more than three centuries. It is known as a table of **common logarithms** and is reproduced in Appendix 2. A common logarithm is simply a power of 10; specifically, **it is the power to which 10 must be raised to give a particular number.*** To understand what this statement means, let us examine Table 2.1.

Numbers such as 0.01, 1, and 100, which can be expressed as integral powers of ten ($0.01 = 10^{-2}$, $1 = 10^{0}$, $100 = 10^{2}$), have integral logarithms

---

*Numbers other than 10 can be used as a base for logarithms (see Section 2.4). As the phrase "common logarithm" implies, 10 is the base most commonly employed. When we use the word "logarithm" without any qualifying adjective, we will mean the "base 10" or "common" logarithm.

Table 2.1   Logarithms of a Few Numbers Between 0.001 and 2000

Number	Logarithm	Number	Logarithm
0.001	−3	0.002	$0.3010 - 3 = -2.6990$
0.01	−2	0.02	$0.3010 - 2 = -1.6990$
0.1	−1	0.2	$0.3010 - 1 = -0.6990$
1	0	2	$0.3010 + 0 = 0.3010$
10	1	20	$0.3010 + 1 = 1.3010$
100	2	200	$0.3010 + 2 = 2.3010$
1000	3	2000	$0.3010 + 3 = 3.3010$

$(-2, 0, 2)$. The number 2, which falls between 1 and 10, is not an integral power of 10. It can be shown, however, that:

$$10^{0.3010} = 2.000$$

Consequently, the logarithm of 2 must be, to four decimal places, 0.3010. Note that, since the number 2 is intermediate between the numbers 1 and 10, its logarithm (0.3010) is intermediate between those of 1 (0) and 10 (1).

Once the logarithm of 2 has been established, logarithms of numbers such as 200 and 0.02 can be immediately written down by taking advantage of the fact that a logarithm is simply an exponent of 10. Thus, we have:

$$
\begin{aligned}
200 &= 2 \times 100 \\
\log 200 &= \log 2 + \log 100 \\
&= 0.3010 + 2 \\
&= 2.3010
\end{aligned}
\qquad
\begin{aligned}
0.02 &= 2 \times 0.01 \\
\log 0.02 &= \log 2 + \log 0.01 \\
&= 0.3010 - 2 \\
&= -1.6990
\end{aligned}
$$

It will be noted from Table 2.1 that numbers greater than 1 have positive logarithms, while numbers less than 1 (but greater than 0) have negative logarithms. That is:

$$\text{if } x > 1, \qquad \text{then } \log x > 0*$$

$$\text{if } 0 < x < 1, \qquad \text{then } \log x < 0$$

As the number approaches zero, its logarithm approaches negative infinity (see Figure 2.1). Numbers less than zero (e.g., -1, -2) cannot be assigned logarithms. It is impossible to obtain a number less than zero by raising 10 to any power whatsoever.

---

*The symbol "$\log x$" will be used to refer to the common (base 10) logarithm of x. In cases where there may be some ambiguity as to the base, we shall write "$\log_{10} x$" to represent the common logarithm.

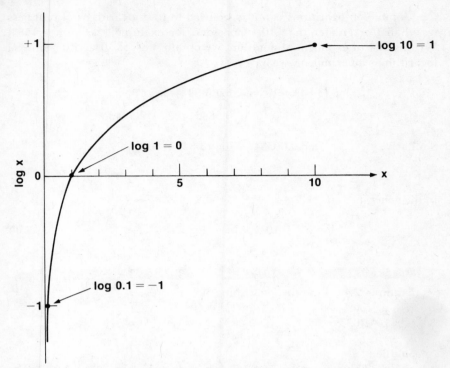

FIGURE 2.1 Numbers greater than 1 have positive logarithms.   Numbers less than 1 have negative logarithms.

2.1   FINDING THE LOGARITHM OF A NUMBER

The four-place table of logarithms given in Appendix 2 allows us to determine directly the logarithm of any three-digit number between 1 and 10. To illustrate, suppose we wish to find the logarithm of 3.84. We follow down the column at the far left of the table until we come to 3.8 and then move across to the column headed "4," reading off the logarithm of 3.84 as 0.5843. Similarly, we would find the logarithm of 3.85 to be 0.5855, that of 3.86 to be 0.5866, and so on.

It is also possible to estimate quite accurately from the table the logarithms of four-digit numbers. Suppose, for example, we want to know the logarithm of 3.845. We know that the logarithms of 3.84 and 3.85 are 0.5843 and 0.5855 respectively. Since the number 3.845 is half way between 3.84 and 3.85, its logarithm should be about half way between 0.5843 and 0.5855, i.e., 0.5849. Expressing our reasoning in mathematical language:

$$\log 3.845 = \log 3.84 + 0.5 (\log 3.85 - \log 3.84)$$
$$= 0.5843 + 0.5 (0.5855 - 0.5843)$$
$$= 0.5843 + 0.5 (0.0012) = 0.5849$$

A table of logarithms can also be used to find logarithms of numbers less than 1 or greater than 10. Consider, for example, the number 384. Writing 384 in exponential notation, we obtain 3.84×10^2. But, since a logarithm is an exponent, we have:

$$\log (3.84 \times 10^2) = \log 3.84 + \log 10^2$$

$$= 0.5843 + 2 = 2.5843$$

Similarly:
$$\log 0.00384 = \log (3.84 \times 10^{-3})$$

$$= 0.5843 - 3 = -2.4157$$

In the general case:

$$\log (C \times 10^n) = n + \log C \qquad (2.1)$$

Example 2.1 Find the logarithms of:
a. 6.02   c. 6.023×10^{23}
b. 6.023   d. 0.6023

Solution
a. The logarithm of 6.02 is found directly from the table; it is 0.7796.
b. The logarithm of 6.023 must be $^3/_{10}$ of the way between the logarithms of 6.02 (0.7796) and 6.03 (0.7803).

$$\log 6.023 = 0.7796 + 0.3(0.7803 - 0.7796)$$

$$= 0.7796 + 0.0002 = 0.7798$$

c. $\log (6.023 \times 10^{23}) = 23 + \log 6.023 = 23 + 0.7798 = 23.7798$
d. $\log 0.6023 = \log (6.023 \times 10^{-1}) = -1 + 0.7798 = -0.2202$

EXERCISES

Find the logarithms of the following numbers:

1. 8.16   5. 56.52
2. 1.03   6. 4.18×10^9
3. 1.652   7. 0.004918
4. 5.841   8. 3.87×10^{-12}

2.2 FINDING THE NUMBER CORRESPONDING TO A GIVEN LOGARITHM

The operation of finding an antilogarithm (number corresponding to a given logarithm) is simply the inverse of finding the logarithm of a number. Experience has shown, however, that students often have difficulty with this operation.

To start with a simple example, let us find the number whose logarithm is 0.4997. To do this, we locate 0.4997 in the body of the table. We note that it falls in the horizontal column labeled "3.1" under the vertical column labeled "6." We deduce that the antilogarithm of 0.4997 is 3.16 (i.e., 3.16 is the number whose logarithm is 0.4997). By the same procedure, we could deduce that the antilogarithm of 0.5011 is 3.17.

Now let us consider the slightly more complicated (and much more common) case, in which the logarithm we are working with does not appear directly in the table. Suppose, for example, we wish to find the antilogarithm of 0.5000. Although we cannot locate this logarithm directly in the table, we can bracket it between 0.4997, the logarithm of 3.16, and 0.5011, the logarithm of 3.17. We deduce that since 0.5000 lies between 0.4997 and 0.5011, its antilogarithm must lie between 3.16 and 3.17. Specifically, 0.5000 is $\frac{3}{14}$ or 0.2 of the way between 0.4997 and 0.5011; we estimate that its antilogarithm is 0.2 of the way between 3.16 and 3.17. Since 3.162 is the number which lies 0.2 of the way from 3.16 to 3.17, it must be the antilogarithm of 0.5000.

Example 2.2   Find the number whose logarithm is 0.8641.

Solution   Scanning the table, we find that:

$$0.8639 = \log 7.31; \quad 0.8645 = \log 7.32$$

Since 0.8641 lies $\frac{2}{6}$, or approximately 0.3 of the way between 0.8639 and 0.8645, it follows that the number must be 0.3 of the way between 7.31 and 7.32. In other words, the antilogarithm of 0.8641 must be approximately 7.313.

The process which we have just described can be applied to logarithms such as 0.5000 or 0.8641, which fall between 0 and 1. Frequently, we need to find numbers corresponding to logarithms which are greater than 1 (e.g., 1.5000, 25.8641) or less than zero (−0.5000, −2.1359). The general principle which we shall use in all such cases is to **rewrite the**

logarithm so that it is in the form of a decimal fraction (mantissa) plus or minus a whole number (characteristic).

To illustrate this procedure, consider the problem of finding the number whose logarithm is 25.8641. We cannot, of course, find this logarithm directly in the table, which is limited to logarithms having values between 0.0000 and 1.0000. Consequently, we rewrite 25.8641 as 0.8641 + 25. Now, we look up the mantissa, 0.8641, in the table, and find that its antilogarithm is 7.313 (see Example 2.2). The antilogarithm of the characteristic, 25, is 10^{25}. It follows that the number we are looking for is 7.313 × 10^{25}. Mathematically:

$$\log (7.313 \times 10^{25}) = \log 7.313 + \log 10^{25}$$
$$= 0.8641 + 25 = 25.8641$$

You will note that what we have done is to *use the mantissa to determine the coefficient of the exponential number; the characteristic simply tells us the exponent.*

If we are dealing with a logarithm which is a negative number, we proceed in an entirely analogous manner. Let us suppose that we wish to find the number whose logarithm is −3.6990. We must rewrite −3.6990 to get it in the form of a decimal fraction, between 0.0000 and 1.0000, minus a whole number. A moment's reflection should convince you that the result will be 0.3010 − 4. That is:

$$-3.6990 = 0.3010 - 4$$

Having passed this hurdle, we proceed as before, noting that the antilogarithm of 0.3010 is 2.00 and that of −4 is 10^{-4}. Consequently, the number we are looking for must be 2.00 × 10^{-4}.

Example 2.3   Find the numbers whose logarithms are:
  a. 6.4771
  b. −1.3980

Solution
  a. antilog 6.4771 = antilog (0.4771 + 6)
    Looking up 0.4771 in the table, we find its antilog to be 3.00; the antilog of 6 is 10^6. Consequently:

$$\text{antilog } (0.4771 + 6) = 3.00 \times 10^6$$

  b. Putting −1.3980 in the standard form, we have:

$$-1.3980 = 0.6020 - 2$$

The antilog of 0.6020 is 4.00; that of -2 is 10^{-2}. Hence, the desired number is 4.00×10^{-2}.

EXERCISES

Find the numbers whose logarithms are:

1. 0.8831
2. 0.9367
3. 1.7435
4. 1.6165

5. -2.4023
6. -2.6195
7. -12.4000
8. 7.6221

2.3 OPERATIONS INVOLVING LOGARITHMS

Since logarithms are exponents, the rules that we derived in Chapter 1 for performing mathematical operations with exponents can be extended to logarithms. The results are summarized in Table 2.2

Table 2.2 Mathematical Operations Involving Exponents and Logarithms

	Exponents	Logarithms
Multiplication	$10^a \times 10^b = 10^{(a+b)}$	$\log (xy) = \log x + \log y$
Division	$10^a/10^b = 10^{(a-b)}$	$\log x/y = \log x - \log y$
Raising to a power	$(10^a)^n = 10^{an}$	$\log x^n = n \log x$
Extracting a root	$(10^a)^{1/n} = 10^{a/n}$	$\log x^{1/n} = \dfrac{1}{n} \log x$

Example 2.4 Using Table 2.2, calculate:

a. $\log (2.061 \times 4.190)$

b. $\log \dfrac{3.160 \times 10^4}{2.082 \times 10^5}$

c. $\log (6.023)^3$

Solution

a. $\log 2.061 = 0.3141$; $\log 4.190 = 0.6222$
$\log (2.061 \times 4.190) = 0.3141 + 0.6222 = 0.9363$

b. $\log 3.160 \times 10^4 = 4.4997$; $\log 2.082 \times 10^5 = 5.3185$
$\log \dfrac{3.160 \times 10^4}{2.082 \times 10^5} = 4.4997 - 5.3185 = 0.1812 - 1$

$\qquad\qquad\qquad\qquad\qquad$ (or -0.8188)

c. $\log (6.023)^3 = 3 \log (6.023) = 3(0.7798) = 2.3394$

The rules listed in Table 2.2 suggest a way of simplifying certain mathematical processes. They enable us to substitute the simpler processes of addition and subtraction for the tedious operations of multiplication and division. Suppose, for example, we wish to multiply 2.061 by 4.190. Following the procedure shown in Example 2.4a, we have:

$$\log(2.061 \times 4.190) = 0.3141 + 0.6222 = 0.9363$$

Using a table of logarithms, we find the antilog of 0.9363 to be 8.636. Consequently, we deduce that:

$$2.061 \times 4.190 = 8.636$$

Similarly, to divide 3.160×10^4 by 2.082×10^5, we have:

$$\log \frac{3.160 \times 10^4}{2.082 \times 10^5} = 4.4997 - 5.3185 = 0.1812 - 1$$

$$\text{antilog } 0.1812 - 1 = 1.518 \times 10^{-1}$$

(Note that in this case, it was more convenient to leave the logarithm in the form $0.1812 - 1$, rather than writing it as -0.8188.)

In practice, most of the multiplications and divisions that we carry out in general chemistry can be accomplished with sufficient accuracy on a slide rule (Chapter 3). Since a slide rule is generally more convenient to use than a table of logarithms, we shall not dwell further on the use of logarithms for multiplication and division. A slide rule can also be used for squaring and cubing numbers and for extracting square or cube roots. However, raising a number to a large power (e.g., 6, or 10) is a tedious operation on a slide rule, as is the process of raising to a very small fractional power (e.g., $1/5$, $1/9$). In situations such as this, the operation is perhaps most conveniently carried out with the aid of a table of logarithms.

Example 2.5   Using logarithms, find $(6.02 \times 10^{23})^{1/8}$

Solution $\log 6.02 \times 10^{23} = 23.7796$

$\log (6.02 \times 10^{23})^{1/8} = 23.7796/8 = 2.9725$

$\text{antilog } 2.9725 \approx 9.39 \times 10^2$

EXERCISES

Making use of a table of logarithms, find:

1. $\log(6.160 \times 10^3)(1.680 \times 10^{-2})$

2. $\log \dfrac{4.983 \times 10^4}{3.172 \times 10^2}$

3. $\log \dfrac{1.684 \times 10^5}{6.480 \times 10^7}$

4. $\log(9.080 \times 10^2)^3$

5. $\log(6.162)^{1/4}$

6. $\log \dfrac{(6.161 \times 10^8)(3.812)^2}{(1.976 \times 10^{-10})(6.180)^{1/2}}$

7. $(7.074)^{19}$

8. $\dfrac{(3.265 \times 10^4)^{1/6}}{(6.010 \times 10^{-2})^5}$

2.4 NATURAL LOGARITHMS

To this point, we have been discussing "common logarithms," i.e., logarithms to the base 10. For calculation purposes, common logarithms are the simplest to work with, since our number system is based on multiples of 10. However, certain of the equations which we use in general chemistry involve a different type of logarithm, taken to the base e* in which:

$$e = 2.718 \cdots$$

Logarithms to the base e are referred to as **natural logarithms,** reflecting the fact that they arise in a natural way in integral calculus (see Chapter 10). To distinguish natural from common logarithms, we use the abbreviation **ln** to stand for a logarithm to the base e:

$$\log_e x \equiv \ln x; \qquad \log_{10} x \equiv \log x$$

Tables of natural logarithms are available, but, in practice, are seldom used. If we wish to find the natural logarithm of a number, we first look

*The base of natural logarithms, e, is defined as the limit approached by the quantity $(1 + 1/n)^n$ as n becomes very large. Thus:

n	$(1 + 1/n)^n$		n	$(1 + 1/n)^n$	
1	2^1	$= 2.000$	10	$(11/10)^{10}$	$= 2.594$
2	$(3/2)^2$	$= 2.250$	100	$(101/100)^{100}$	$= 2.705$
3	$(4/3)^3$	$= 2.370$	∞		$= 2.718 \cdots$

up the common logarithm, and then make use of the equation:*

$$\ln x = 2.303 \log x \qquad (2.2)$$

Example 2.6   Find the natural logarithm of:
   a. 2.280×10^3
   b. 9.831×10^{-2}

Solution   In both cases, we first find the base 10 log and then multiply by 2.303.
   a. $\log(2.280 \times 10^3) = 3.3579$;
      $\ln(2.280 \times 10^3) = (2.303)(3.3579) = 7.733$
   b. $\log(9.831 \times 10^{-2}) = 0.9926 - 2 = -1.0074$;
      $\ln(9.831 \times 10^{-2}) = (2.303)(-1.0074) = -2.320$

Occasionally, we need to evaluate expressions in which the base of natural logarithms, e, is raised to a power (e.g., $e^{0.2500}$). This can readily be accomplished with the aid of Equation 2.2 and a table of common logarithms. The procedure is indicated in Example 2.7.

Example 2.7   Find the numerical value of $e^{0.2500}$.

Solution   If we let $x = e^{0.2500}$

$$\ln x = 0.2500$$

$$2.303 \log x = 0.2500; \quad \log x = 0.2500/2.303 = 0.1086$$

Taking antilogarithms: $x = 1.284$

---

*To derive this equation, let $x = e^y$. Taking:

natural logs: $\ln x = y$

common logs: $\log_{10} x = y \log_{10} e$

Dividing: $\dfrac{\ln x}{\log_{10} x} = \dfrac{1}{\log_{10} e} = \dfrac{1}{\log_{10} 2.718} = \dfrac{1}{0.4343} = 2.303$

EXERCISES

Evaluate:
1. ln 6.023.
2. ln (2.02×10^3).
3. ln (6.18×10^{-5}).
4. The base 10 log of the number whose natural logarithm is one.
5. The base 10 log of the number whose natural logarithm is 6.190.
6. $e^{22.96}$
7. $e^{-1.000}$

PROBLEMS

In working these problems, the following relationships will be useful:

$$pH \equiv -\log(\text{conc. } H^+)$$

$$T = \text{temperature in } °K = \text{temperature in } °C + 273°$$

$$R = \text{gas constant} = 1.99 \text{ cal/}°K$$

2.1 Calculate the pH of solutions which have the following concentrations of H^+:

    a. 1.0×10^{-6} c. 3.0×10^{-9}

    b. 2.61×10^{-2} d. 6.0

2.2 The concentration of H^+ in solution A is 100 times as great as that in solution B. What is the difference in pH between the two solutions? Which solution has the greater pH?

2.3 In a 0.10 M solution of acetic acid, the concentration of H^+ is given by the expression:

$$(\text{conc. } H^+)^2 = 1.80 \times 10^{-6}$$

What is the pH of this solution?

2.4 Calculate the concentration of H^+ in solutions of the following pH:

    a. 4.0 c. 3.14

    b. 12.60 d. -1.0

2.5   In a solution saturated with hydrogen sulfide, the following relation holds:

$$(\text{conc. } H^+)^2 \times (\text{conc. } S^{2-}) = 1 \times 10^{-23}$$

What is the concentration of S^{2-} in a solution of this type which has a pH of 4.0?

2.6   In any water solution at 25°C:

$$(\text{conc. } H^+) \times (\text{conc. } OH^-) = 1.0 \times 10^{-14}$$

The term pOH is defined as:

$$pOH \equiv -\log (\text{conc. } OH^-)$$

Making use of either or both of these relations, calculate:
   a. The concentration of OH^- in a solution of pH 6.0.
   b. The pOH of a solution in which the concentration of OH^- is 1×10^{-4}.
   c. The pOH of a solution in which the concentration of H^+ is 2.0×10^{-3}.

2.7   The standard free energy change of a reaction, $\Delta G°$ (in calories), is related to the equilibrium constant, K, by the expression:

$$\Delta G° = -2.303 \; RT \log K$$

   a. If $\Delta G° = 0$, $K =$ ———— .
       If $\Delta G° < 0$, K is ———— than one.
       If $\Delta G° > 0$, K is ———— than one.
   b. Calculate $\Delta G°$ at 298°K for a reaction for which $K = 1.60 \times 10^{-4}$.
   c. Calculate K at 1000°K for a reaction for which $\Delta G° = -12,000$ cal.

2.8   The potential, E, for the reduction of Zn^{2+} is given by the equation:

$$E = -0.76 - 0.030 \log [1/(\text{conc. } Zn^{2+})]$$

   a. Calculate E when the concentration of Zn^{2+} is 1.0×10^{-8}.
   b. Calculate the concentration of Zn^{2+} when $E = -1.52$.

2.9 The equation for the rate of radioactive decay is:

$$\log \frac{X_0}{X} = \frac{kt}{2.303}$$

where X_0 is the original amount of radioactive material, X is the amount remaining after time t, and k is the rate constant.

    a. If $k = 2.00 \times 10^{-3}$/min and $X_0 = 0.0100$ g, what is X when $t = 50.0$ min?

    b. How long will it take for the amount of radioactive material to drop from 1.0 g to 0.10 g if $k = 1.0 \times 10^{-2}$/sec?

    c. Show that the time required for one half of the radioactive material to decay must be $0.693/k$.

2.10 The rate constant, k, for a reaction can be expressed as a function of temperature by the relation:

$$\ln k = \frac{-E_a}{RT} + B$$

where E_a is the energy of activation and B is a constant. Calculate k at 25°C for a reaction for which $B = 2.10$ and $E_a = 1.0 \times 10^4$ cal.

2.11 The vapor pressure of water is given as a function of temperature by the relation:

$$\log \frac{P_2}{P_1} = \frac{10{,}200 \, (T_2 - T_1)}{2.303 \, RT_2T_1}$$

where P_2 and P_1 are the vapor pressures at temperatures T_2 and T_1 respectively. The vapor pressure of water at 100°C is 760 mm Hg.

    a. What is the vapor pressure of water at 25°C (298°K)?

    b. At what temperature will the vapor pressure of water be 1.20×10^3 mm Hg?

2.12 The fraction, f, of molecules having an energy equal to or greater than the activation energy, E_a, is given by the equation:

$$f = e^{-E_a/RT}$$

What is f when $E_a = 1.00 \times 10^4$ cal., $T = 298$°K?

2.13 The Boltzmann equation for the distribution of molecules among two energy levels is:

$$\frac{n_2}{n_1} = e^{(E_1 - E_2)/RT}$$

where n_2 and n_1 are the numbers of molecules in the levels whose energies are E_2 and E_1, respectively. Calculate the ratio n_2/n_1 when $E_1 = 0$ and

a. $E_2 = 0$, $T = 300°K$.
b. $E_2 = 1000$ cal., $T = 300°K$.
c. $E_2 = 1000$ cal., $T = 600°K$.

CHAPTER 3

THE SLIDE RULE

We saw in Chapter 2 the method by which a table of logarithms can be used to multiply, divide, raise to a power, or extract a root. These operations can be carried out more rapidly, with some loss of accuracy, by means of a slide rule. A simple 10-inch rule, selling for about $2, is adequate for most of the calculations that we carry out in general chemistry. The time saved on homework, laboratory calculations, and examinations is well worth the cost of the rule.

It is important to understand the principle upon which the slide rule is based. When we carry out the processes of multiplication or division on a slide rule, what we are really doing is adding or subtracting logarithms. To understand how this comes about, let us examine the scales on the slide rule labeled C and D, which are identical except that the C scale is located on the movable slide and the D scale is on the stationary part of the rule. On these scales, the positions of the numbers are determined by the values of their logarithms. For example, the number 2 is located about 30 per cent of the way from the left *index* (1) to the right *index* (10). This reflects the fact that the logarithm of 2 (0.3010) lies 30.10 per cent of the way between the logarithm of 1 (0.0000) and that of 10 (1.0000). Similarly, the number 3 (log 3 = 0.4771) falls at a point a little less than 50 per cent of the distance between 1 and 10.

Consider now the slide rule setting shown in Figure 3.1. Here, the left index of the C scale is placed directly above the number 2 on the D scale. If we move across the scales, we note that 3 on the C scale is directly above 6 on the D scale. Clearly, by this simple process we have multiplied 2 × 3 to obtain an answer of 6. What we have really done is to *add* two distances. One of these (3.01 inches) is proportional to log 2 (0.3010); the other (4.77 inches) is proportional to log 3 (0.4771). The total distance, 7.78 inches, is proportional to the log of 6, 0.7781. In other words, we multi-

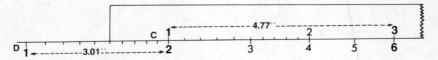

FIGURE 3.1. The positions of numbers on the C and D scales are determined by their logarithms.

log 2 = 0.301	Distance 1 → 2	= 3.01 in.
log 3 = 0.477	Distance 1 → 3	= 4.77 in.
log 6 = 0.778	Distance 1 → 6	= 7.78 in.

plied two numbers by, in effect, adding their logarithms. In an entirely analogous manner, it is possible to carry out divisions by subtracting distances proportional to the logarithms of the numbers involved.

3.1 LOCATING NUMBERS

To illustrate how numbers can be located on the various scales of the slide rule, let us examine the D scale. The large digits (1, 2, 3, 4 · · ·) represent the integers between 1 and 10. Note that the spaces between these integers become smaller as we move from left to right. This gradual compression of the scale is a consequence of its logarithmic nature. The distance between the numbers 1 and 2 is about 3 inches (log 2 − log 1 = 0.3010), while that between 9 and 10 is less than half an inch (log 10 − log 9 = 0.0458).

We shall now consider how to find three-digit numbers in various portions of the D scale (Example 3.1). The positions of these numbers are indicated by dotted vertical lines in Figure 3.2.

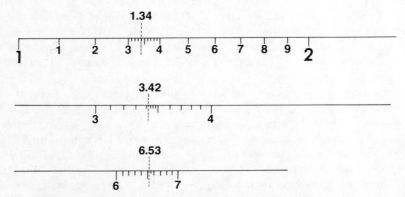

FIGURE 3.2. Location of the numbers 1.34, 3.42, and 6.53 on the D (or C) scale.

Example 3.1   Locate the following numbers on the D scale:
- a. 1.34
- b. 3.42
- c. 6.53

Solution

- a. Focusing our attention on the portion of the D scale between the large digits 1 and 2, we recognize ten major divisions, each labeled with a small digit (1, 2, 3, 4) corresponding to the numbers 1.1, 1.2, 1.3, 1.4 · · · . Between the
  numbers 1.3 and 1.4, there are ten small divisions. The
  number 1.34 must then fall four small divisions to the right
  of 1.3.
- b. Here, we note that the ten major divisions between the
  large digits 3 and 4 are not numbered. To find the number
  3.4, we move four major divisions to the right of the digit 3.
  To locate 3.42, we note that there are five small divisions
  between 3.4 and 3.5, corresponding to 3.42, 3.44, 3.46,
  3.48, and 3.50. The number 3.42 must then be located one
  small division to the right of 3.4.
- c. We first locate the number 6.5, five major divisions to the
  right of 6. Between 6.5 and 6.6, there are two small divisions, which must correspond to 6.55 and 6.60. The number 6.53 is located a little more than half way (viz., $^3/_5$ of
  the way) between 6.50 and 6.55.

EXERCISES

1. Set the following numbers on the D scale:
   - a. 2.02
   - b. 1.68
   - c. 3.42
   - d. 3.69
   - e. 5.15
   - f. 5.74
2. Locate the numbers in part (1) on the A scale, between the left and
   center index. Note that the A scale consists of two segments, one
   running from 1 to 10 and the other from 10 to 100.
3. Set 8 on the C scale directly above 4 on the D scale. Read the
   quotient, 2, on the C scale directly above the left index of the D
   scale. Why does this procedure enable you to carry out the process
   of division? Explain in terms of the logarithmic nature of the
   scales.
4. It is reasonable to suppose that the error which a beginner makes
   in setting a number on the slide rule is equivalent to the smallest

division on the D scale, which could be the distance between 1.99 and 2.00 or, alternatively, that between 9.95 and 10.00. (Note that these divisions are of about the same size.) What percentage of error does this represent?

3.2 MULTIPLICATION AND DIVISION

To carry out operations involving only multiplication and division, we use the C and D scales. Two simple multiplications are illustrated in Example 3.2.

Example 3.2 Multiply:
   a. 2.12 × 4.35
   b. 5.05 × 4.33 (see Fig. 3.3)

Solution
   a. We first set the 1 at the left of the C scale (the left index) directly above 2.12 on the D scale. Now, move the transparent plastic *cursor* so that its hairline lines up exactly with 4.35 on the C scale. Read the answer, approximately 9.22, at the hairline on the D scale.
   b. If we attempt to repeat the procedure of part (a), we find that the number 4.33 on the C scale falls beyond the end of the D scale. To get around this difficulty, we set the *right index* of the C scale above 5.05 on the D scale. Now, move the cursor so that its hairline coincides with 4.33 on the C scale. Directly beneath, on the D scale, we find that the hairline falls about half way across the space between 2.18 and 2.20. We might estimate its position to be 2.19. Knowing that the product must be approximately 20 (5 × 4 = 20), we take 21.9 to be our answer.

In general, to multiply one number by another, we:
1. **Set the appropriate index of the C scale (left or right) directly above the first number on the D scale.**
2. **Move the cursor so that its hairline falls on the second number, on the C scale.**
3. **Read the answer directly beneath the hairline on the D scale.**
The manipulation involved in Example 3.2 part (b) illustrates the fact that the decimal point in the answer must be set independently of the slide

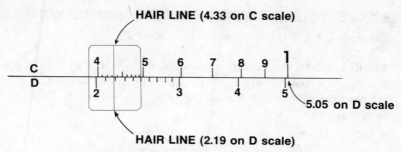

FIGURE 3.3. Multiplication: 5.05 × 4.33 = 21.9.

rule operation. Locating the decimal point, or the appropriate power of 10, is facilitated by expressing the numbers in exponential notation. Suppose, for example, we wish to multiply 0.0212 by 4350. Setting the left index of the C scale above 2.12 on the D scale and moving the cursor until its hairline coincides with 4.35 on the C scale, we read the answer beneath the hairline on the D scale as 9.22. To obtain the power of 10 in the answer, we write the numbers in exponential notation:

$$0.0212 \times 4350 = 2.12 \times 10^{-2} \times 4.35 \times 10^{3}$$

$$= (2.12 \times 4.35) \times (10^{-2} \times 10^{3}) \approx 9 - 10^{1}$$

We deduce that the product must be 9.22×10^{1}, or 92.2. Similarly:

$$0.212 \times 43.5 = 2.12 \times 10^{-1} \times 4.35 \times 10^{1} = 9.22$$

$$212 \times 0.00435 = 2.12 \times 10^{2} \times 4.35 \times 10^{-3} = 9.22 \times 10^{-1} = 0.922$$

Note that in each case the slide rule operations are identical. We use the rules of exponents to determine the power of 10, or the decimal point, in the answer.

Since the process of division is the inverse of multiplication, it can be accomplished on the slide rule by performing in reverse order the manipulations involved in multiplication. Specifically, to divide one number by another, we:

1. **Move the cursor so that its hairline falls directly over the numerator on the D scale.**
2. **Move the slide until the denominator, on the C scale, falls beneath the hairline of the cursor.**
3. **Read the answer on the D scale, directly below the index of the C scale.**

Example 3.3   Divide 6.05×10^3 by 9.50×10^{-4}.

Solution   Using the cursor, we line up 6.05 on the D scale with 9.50 on the C scale. Directly below the right index of the C scale, we read the number "6.37" on the D scale. To decide upon the proper power of 10, we note that:

$$\frac{6.05 \times 10^3}{9.50 \times 10^{-4}} = \frac{6.05 \times 10^7}{9.50} \approx 0.6 \times 10^7$$

It follows that our answer must be 0.637×10^7, or, in standard exponential notation, 6.37×10^6.

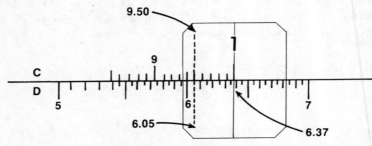

FIGURE 3.4.   Division: $6.05/9.50 = 0.637$.

Many times, in working problems, we are required to carry out a series of successive multiplications and divisions. In doing this, considerable time can be saved by alternating the processes, first dividing, then multiplying, then dividing, and so on.

Example 3.4 $\dfrac{(6.02 \times 10^{23})(1.88 \times 10^{-8})(3.14)}{(1.24 \times 10^8)(5.10 \times 10^4)} = ?$

Solution   Let us first divide 6.02 by 1.24, then multiply by 1.88, divide by 5.10, and, finally, multiply by 3.14.

  a. To divide 6.02 by 1.24, use the cursor to line up 6.02 on the D scale with 1.24 on the C scale. The quotient, which need not be recorded, can be found on the D scale below the left index of the C scale. Note that the rule is now in position to carry out a multiplication.

    b. To multiply by 1.88, move the cursor so that its hairline falls at 1.88 on the C scale. The product will appear on the D scale, beneath the hairline.

    c. To divide by 5.10, leave the cursor where it is and move the slide until 5.10 on the C scale lines up with the hairline.

    d. To multiply by 3.14, move the cursor so that the hairline falls at 3.14 on the C scale. Read "5.62" on the D scale.

To obtain the proper exponent in the answer, we approximate as follows:

$$\frac{(6 \times 10^{23})(2 \times 10^{-8})(3)}{(1 \times 10^{8})(5 \times 10^{4})} = \frac{36 \times 10^{15}}{5 \times 10^{12}} \approx 7 \times 10^{3}$$

Clearly, the answer must be 5.62×10^{3}.

Try carrying out all the multiplications first and then performing the divisions! You will find that several extra manipulations are involved. This should convince you that the recommended procedure of alternating multiplications with divisions has considerable merit.

EXERCISES

Perform the indicated operations on the slide rule.

1. $(6.41 \times 10^{2}) \times (1.39 \times 10^{-4})$
2. 8.54×2.90
3. $\dfrac{7.12 \times 10^{4}}{6.10 \times 10^{-7}}$
4. $1.28/5.90$
5. $\dfrac{(6.19 \times 10^{8}) \times (3.20)}{2.91}$
6. $\dfrac{(1.86) \times (3.95)}{(2.87) \times (4.19)}$
7. $\dfrac{(5.82 \times 10^{7}) \times (4.19 \times 10^{8}) \times (3.14)}{(1.88 \times 10^{14}) \times (6.12 \times 10^{6})}$

3.3   SQUARES AND SQUARE ROOTS

To square a number or extract its square root, we make use of the A and D scales. You will note that the A scale is made up of two ranges, each running from 1 to 1. It is convenient to think of the index at the center of the A scale as representing the number 10 and that at the far right as representing the number 100. From this point of view, the range at the left runs from 1 to 10, while that at the right runs from 10 to 100.

Squaring a number on the slide rule is a very simple procedure. All we do is to **set the hairline of the cursor above the number on the D scale and read its square on the A scale.** Thus, we find (Figure 3.5) that $2^2 = 4, 3^2 = 9, (2.14)^2 = 4.58$.

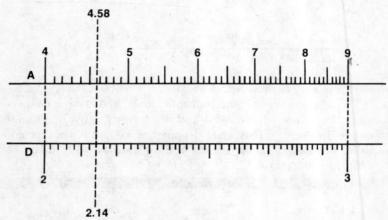

FIGURE 3.5. Squares of numbers: $2^2 = 4;$ $(2.14)^2 = 4.58;$ $3^2 = 9.$

To find $(2.14 \times 10^{-6})^2$, we use the rules of exponents (Chapter 1).

$$(2.14 \times 10^{-6})^2 = (2.14)^2 \times (10^{-6})^2 = 4.58 \times 10^{-12}$$

Extracting the square root of a number is a slightly more complicated manipulation. In principle, all we have to do is to reverse the process involved in squaring the number. That is, we **set the hairline of the cursor above the number on the A scale and read the square root on the D scale.** In this way, we deduce that:

$$(4)^{1/2} = 2$$
$$(12.0)^{1/2} = 3.46$$
$$(50.0)^{1/2} = 7.07$$

A difficulty arises if the number whose square root we are asked to find does not fall between 1 and 100. In this case, **we rewrite the number in exponential form so that the coefficient is a number between 1 and 100 and the exponent is an even power** (e.g., $-2, 0, 2, 4 \cdots$).

Example 3.5 Find:

    a. $(12,600)^{1/2}$

    b. $(1.86 \times 10^5)^{1/2}$

Solution

    a. $(12,600)^{1/2} = (1.26 \times 10^4)^{1/2} = (1.26)^{1/2} \times 10^2$

Setting the hairline of the cursor at 1.26 on the A scale, we read 1.13 on the D scale. Our answer is 1.13×10^2, or 113.

    b. Here, we must first convert the exponent to an even number. To do this, we multiply 1.86 by 10 and divide 10^5 by 10:

$$(1.86 \times 10^5)^{1/2} = (18.6 \times 10^4)^{1/2} = (18.6)^{1/2} \times 10^2$$

To find the square root of 18.6, we move the hairline of the cursor to 18.6 on the A scale (note that 18.6 falls beyond 10, on the right half of the A scale). We read 4.31 on the D scale; our answer must then be 4.31×10^2.

EXERCISES

Find

1. $(1.69)^2$
2. $(6.12 \times 10^{-3})^2$
3. $(1.60)^{1/2}$
4. $(87.0)^{1/2}$

5. $(2.02 \times 10^2)^{1/2}$
6. $(3.86 \times 10^7)^{1/2}$
7. $(2.98 \times 10^{-5})^{1/2}$

3.4 CUBES AND CUBE ROOTS

To cube a number or extract its cube root, we use the K scale in much the same way that the A scale is used to obtain squares or square roots. Notice that the K scale is divided into three segments, which may be thought of as running from 1 to 10 (left), 10 to 100 (center), and 100 to 1000 (right). To find the cube of a number, we **set the hairline of the cursor over that number on the D scale and read the answer under the hairline on the K scale.** Thus, we find that:

$$2^3 = 8; \quad (3.0)^3 = 27; \quad (8.00)^3 = 512$$
$$(3.0 \times 10^{-3})^3 = (3.0)^3 \times (10^{-3})^3 = 27 \times 10^{-9}$$

Cube roots of numbers between 1 and 1000 are readily found by **setting the hairline of the cursor above the number on the K scale and**

reading the answer beneath the hairline on the D scale. We suggest that you use your slide rule to confirm that:

$$(3.00)^{1/3} = 1.44$$
$$(30.0)^{1/3} = 3.11$$
$$(300)^{1/3} = 6.69$$

To extract the cube root of a number which does not fall between 1 and 1000, it is necessary to rewrite it in exponential form so that the coefficient is a number between 1 and 1000 and the exponent is evenly divisible by three.

Example 3.6 Find:
  a. $(0.00190)^{1/3}$
  b. $(4.68 \times 10^{-5})^{1/3}$

Solution
  a. $(0.00190)^{1/3} = (1.90 \times 10^{-3})^{1/3} = (1.90)^{1/3} \times 10^{-1}$
Below 1.90 on the K scale, we read 1.24 on the D scale. Our answer must be 1.24×10^{-1}, or 0.124.
  b. We must first rewrite the number so that the exponent, upon division by 3, will give a whole number. To obtain the desired form, we multiply the coefficient by 10 and divide the exponent by 10.

$$(4.68 \times 10^{-5})^{1/3} = (46.8 \times 10^{-6})^{1/3} = (46.8)^{1/3} \times 10^{-2}$$

Setting 46.8 on the K scale (between the second 4 and 5), we read its cube root on the D scale as 3.60. Hence, our answer must be 3.60×10^{-2}.

EXERCISES

1. $(6.08)^3$
2. $(1.40 \times 10^2)^3$
3. $(5.95 \times 10^{-4})^3$
4. $(12.9)^{1/3}$
5. $(268)^{1/3}$
6. $(1.96)^{1/3}$
7. $(1.37 \times 10^4)^{1/3}$
8. $(5.92 \times 10^{-4})^{1/3}$

3.5 LOGARITHMS

The only linear scale on the slide rule, where the numbers are spaced at equal intervals, is the L scale. This scale gives the logarithms of the numbers between 1 and 10. To find the logarithm of a number in this range, we set the hairline of the cursor above the number on the D scale and read its logarithm on the L scale. In this way, we find that the logarithm of 2 is, to three digits, 0.301, that of 3 is 0.477, and so on.

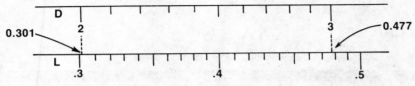

FIGURE 3.6. Logarithms of numbers: log 2 = 0.301;　log 3 = 0.477.

If the number whose logarithm we require is less than 1 or greater than 10, we proceed as described in Chapter 2. That is, we write the number in standard exponential notation, use the L scale to find the mantissa, and take the exponent to be the characteristic.

Example 3.7　Find the logarithm of 3.70×10^{-5}.

Solution　$\log(3.70 \times 10^{-5}) = \log 3.70 + \log 10^{-5} = \log 3.70 - 5$
Setting 3.70 on the D scale, we read its logarithm, on the L scale, to be 0.568. Hence, $\log(3.70 \times 10^{-5}) = 0.568 - 5 = -4.432$

We see from this example that the L scale is equivalent to a three-place table of logarithms. All of the operations which were described in Chapter 2 in connection with a table of logarithms can be carried out on a slide rule, using the L and D scales. We can, for example, use it to find antilogarithms.

Example 3.8　Find the numbers whose logarithms are:
    a. 0.380
    b. 3.100
    c. −2.900

Solution
    a. Set the hairline of the cursor over 0.380 on the L scale; read the number, 2.40, on the D scale.
    b. Set the hairline over the mantissa, 0.100, on the L scale. The number whose log is 0.100 is read from the D scale to

be 1.26. The number whose log is 3 is 10^3. Therefore:

$$\text{antilog } 3.100 = 1.26 \times 10^3$$

c. First, rewrite the number (recall the discussion in Chapter 2) as:

$$0.100 - 3$$

$$\text{antilog } (0.100 - 3) = 1.26 \times 10^{-3}$$

EXERCISES

Find, using the slide rule:

1. log 6.14
2. log (1.29×10^5)
3. log (5.84×10^{-4})
4. ln 2.82
5. antilog 0.462
6. antilog 1.250
7. antilog (-5.400)

3.6   OTHER USES OF THE SLIDE RULE

The slide rule operations performed most frequently in general chemistry have been described in Sections 3.2 through 3.5. Depending on the number of scales available (and the price!), it is possible to use the slide rule to carry out a variety of other operations. Even the simplest slide rule has scales (labelled S and T) which allow us to obtain trigonometric functions (sine, tangent). A discussion of this application of the slide rule will be found in Chapter 8.

From time to time, we may have occasion to find the reciprocal of a number (the reciprocal of a number, x, is defined as $1/x$). This can, of course, be done by treating it as an ordinary division and making use of the C and D scales. Where a series of reciprocals is required, it is somewhat more convenient to make use of the C and C1 scales. Notice that the C1 scale amounts to an inverse C scale; the numbers on the C1 scale start at 10 on the far left and decrease to 1 on the far right. To find the reciprocal of a number, we set the hairline of the cursor over that number on the C scale and read its reciprocal at the hairline on the C1 scale. In this way, you can verify that:

$$1/2 = 0.500; \quad 1/6 = 0.167; \quad 1/3.16 = 0.316$$

The C1 scale can also be used in conjunction with the D scale to carry out the processes of multiplication and division. In both cases, the manipulation is somewhat more rapid than is the case when the C and D scales are used (Section 3.2).

To multiply 9 by 12, using the C1 and D scales, use the cursor to line up 9 on the C1 scale with 12 on the D scale. Read the product, 108, on the D scale, directly below the index of the C1 (or C) scale. Note that when multiplications are carried out in this way, it is never necessary to shift indices to bring the answer onto the scale.

To divide 7.0 by any desired number (e.g., 5.0, 3.5, 2.0), set the index of the C1 (or C) scale to 7 on the D scale. Move the cursor to the number by which you wish to divide, on the C1 scale. Read the answer at the hair-line on the D scale. In this way, we find that:

$$7.0/1.4 = 5.0$$

$$7.0/2.0 = 3.5$$

$$7.0/3.5 = 2.0$$

Note the advantage of this method of performing divisions when you are asked to divide a given number, such as 7.0, by a series of numbers (1.4, 2.0, 3.5 · · ·).

PROBLEMS

The following problems are designed to review material on exponents (Chapter 1) and logarithms (Chapter 2), in addition to improving your speed and accuracy in using the slide rule.

Make use of the slide rule to obtain numerical answers to each problem.

3.1 $(5.49 \times 10^2)(6.02 \times 10^{-8})$

3.2 $\dfrac{4.91 \times 10^{-2}}{6.84 \times 10^3}$

3.3 $(3.08 \times 10^7)^2$

3.4 $\log(2.41 \times 10^5)$

3.5 $(7.28 \times 10^{-6})^{1/3}$

3.6 antilog 0.649

3.7 $(5.00 \times 10^2)^{1/2}$

3.8 $(1.29 \times 10^{-22})^3$

3.9 $1/(1.98 \times 10^{-4})$

3.10 $\dfrac{(1.28 \times 10^7)(3.14)}{6.98 \times 10^3}$

3.11 $2.71(3.08 \times 10^{-3})^{1/2}$

3.12 $\dfrac{(6.47 \times 10^{-2})(8.19 \times 10^{7})}{(3.25 \times 10^{4})(4.21 \times 10^{-6})}$

3.13 antilog 2.490

3.14 3.23 log 6.00

3.15 $\dfrac{(6.19 \times 10^{4})(2.00 \times 10^{-3})^{2}}{9.19 \times 10^{4}}$

3.16 $\dfrac{6.18 (4.14 \times 10^{-5})^{1/3}}{2.00}$

3.17 $\dfrac{(9.62 \times 10^{-7})(4.08 \times 10^{4})(1.72 \times 10^{5})}{(4.43 \times 10^{2})(5.09 \times 10^{-4})}$

3.18 $e^{1.80}$

3.19 1.26×10^{3} (antilog 4.10)

3.20 $\dfrac{(2.19 \times 10^{-2})^{1/2}(4.80 \times 10^{3})^{2}}{(6.18 \times 10^{-5})^{1/3}}$

SIGNIFICANT FIGURES

The numbers that we deal with in general chemistry can be divided into two broad categories. Some numbers are, by their nature, **exact**. When we say that there are *two* crucibles or *five* beakers in a laboratory locker, we indicate exactly how many such items we have. Other numbers are **inexact;** when we refer to a "100 ml beaker," we do not imply that it has a volume of *exactly* 100 milliliters. If we fill the beaker with water we may find that it holds as little as 90 ml or as much as 110 ml.

An important type of exact number is one that is used to relate units within the same measuring system. When we say that:

$$1 \text{ ft } = 12 \text{ in}$$

$$1 \text{ Å } = 1 \times 10^{-8} \text{ cm}$$

the numbers 12 and 1×10^{-8} are exact; they arise because of the way the inch and the Angstrom (Å) are defined. On the other hand, the numbers which are used to relate units in two different measuring systems are inexact. For example, when we say that

$$1 \text{ in } = 2.54 \text{ cm}$$

we do not mean that one inch is exactly equal to 2.54 centimeters. The two units, the centimeter and the inch, are defined independently of one another; there are *approximately* 2.54 centimeters in one inch.

Numbers which arise from experimental measurements are always inexact. The uncertainty in a measurement depends upon the skill of the experimentalist and the sensitivity of the instrument he uses. If we were asked to weigh out a sample of sodium chloride on a triple beam balance, we might be able to establish its weight to within 0.01 g. Perhaps we would report that it weighed:

$$2.65 \pm 0.01 \text{ g}$$

If we needed to know the mass more accurately, we could use a balance with a sensitivity of 0.001 g, in which case we might find that the sample weighed:

$$2.652 \pm 0.001 \text{ g}$$

Analytical balances capable of weighing to ± 0.0001 g are readily available; using such a balance, we might report the mass to be:

$$2.6518 \pm 0.0001 \text{ g}$$

From the standpoint of a person who is trying to estimate the validity of an experiment or to repeat it in another laboratory, it is important that we specify the uncertainty associated with a measurement. One way to do this is to use the $\pm$ notation just shown. Frequently, we omit the ± 0.01, ± 0.001, and so on, and simply report:

$$2.65 \text{ g}; \qquad 2.652 \text{ g}; \qquad 2.6518 \text{ g}$$

with the understanding that there is an *uncertainty of one unit in the last digit.* When we say that a sample of sodium chloride weighs "2.65 g," we imply that its mass is between 2.64 g and 2.66 g.

The precision of measurements such as this can also be described in terms of the number of **significant figures.** We say that in "2.65 g" there are three significant figures; each of the three digits is experimentally significant. The masses "2.652 g" and "2.6518 g" are quoted to 4 and 5 significant figures respectively, implying successively higher degrees of precision.

We shall find the concept of significant figures a very useful one in expressing the reproducibility, not only of individual measurements, but also of calculated quantities based upon such measurements. In this chapter, we shall be concerned with the use of significant figures as a measure of experimental precision. Later, in Chapter 11, we will present a more sophisticated treatment of this topic based on statistical principles.

4.1 COUNTING SIGNIFICANT FIGURES

Frequently, we are faced with the problem of deciding how many significant figures there are in a number which arises from an experiment performed by another person. In many cases, there is no ambiguity. When we find in a table the atomic weight of calcium listed as 40.08, we trust that it is known to four significant figures.

When either the first or the last digit in an experimental quantity is zero, the number of significant figures may not be immediately obvious. Common sense is an invaluable guide here. When we find the atomic weight of krypton listed as 83.80, it should be clear that it is known to four significant figures. If the zero were not significant, there would be no reason for including it. Writing "83.80" implies that the true value of the atomic weight of krypton lies between 83.79 and 83.81.

In a slightly less obvious case, suppose we are told that the volume of a certain object is 0.02461 liters. Is the zero immediately to the right of the decimal point significant? A moment's reflection should convince you that it is not; in this case, *the zero is used simply to fix the position of the decimal point.* To make this conclusion more obvious, suppose we were to express the volume in milliliters rather than liters. Since 1 liter = 1000 ml, the volume would now be 24.61 ml, with four significant figures clearly indicated. Since we cannot change the precision of a measurement by changing the units in which it is reported, there must also be four significant figures in the quantity 0.02461 liters.

Suppose we are told to weigh out 500 g of sodium chloride for a certain experiment. To how many significant figures are we to make this measurement: one? two? three? Unfortunately, we cannot answer this question unless we are able to read the mind of the person who wrote the directions for the experiment. He might have meant to weigh out roughly a quantity between 400 and 600 g, i.e., 500 ±100 g. If so, there is one significant figure. On the other hand, we may be expected to weigh to ±10 g (2 significant figures). Then again, perhaps we are supposed to weigh to the nearest gram, in which case the number "500" has three significant figures. About all we can do in cases like this is to wish that the person giving the directions had expressed the mass in standard exponential notation as:

$$5 \times 10^2 \text{ g} \qquad (1 \text{ significant figure})$$

or: $\qquad 5.0 \times 10^2 \text{ g} \qquad (2 \text{ significant figures})$

or: $\qquad 5.00 \times 10^2 \text{ g} \qquad (3 \text{ significant figures})$

in which case there would have been no ambiguity.

Example 4.1   How many significant figures are there in:
a. 1.204×10^{-2} g   c. 0.00281 g
b. 3.160×10^8 Å   d. 810 ml

Solution

    a. All digits are significant; 4.

    b. All digits are significant; 4.

    c. Three. The zeros serve only to fix the position of the decimal point.

    d. Perhaps 2 (8.1×10^2 ml) or 3 (8.10×10^2 ml).

Occasionally, you may be asked a question such as, "How many grams of carbon dioxide can be produced from one gram of carbon?" In a case such as this, where the number is written out, you are to assume that it is exact. Essentially, you are being asked a rhetorical question, "How much carbon dioxide could be produced from one gram of carbon if it could be weighted out exactly?" For calculation purposes, we assume an infinite number of significant figures in the number "one," just as we consider the number "12" in the defining equation:

$$1 \text{ ft} = 12 \text{ in}$$

to have an infinite number of significant figures.

EXERCISES

Give the number of significant figures in:

1. 12.82 lit
2. 3.19×10^{15} atoms
3. 4.300×10^{-6} cm
4. 0.00641 g
5. 8.2354×10^{-9} m
6. 0.0559 g
7. 2.92×10^2 g
8. 4.1 lit
9. 0.0002 cm
10. 450 g

4.2 MULTIPLICATION AND DIVISION

Most of the quantities that we measure in the laboratory are not end results in themselves; they are used to calculate other quantities. We may, for example, multiply the length of a piece of tin foil by its width to determine its area. In another case, we may divide the mass of a sample by its volume to determine its density.

When we multiply or divide two experimental quantities, both of which are inexact, the product or quotient is inexact. The question arises as to how great an error is introduced by these operations. To answer this question, it will be helpful to work through a specific example. Let us suppose that we have measured the mass and volume of a sample and found them to be 5.80 ±0.01 g and 2.6 ±0.1 ml, respectively. The density, found by dividing mass by volume, should be approximately (5.80/2.6) g/ml. However, taking into account the uncertainties in the experimental quantities, we realize that the density might be as large as (5.81/2.5) g/ml or as small as (5.79/2.7) g/ml. Carrying out these divisions, we obtain:

$$
\begin{array}{ccc}
\underline{2.32\ldots} & \underline{2.23\ldots} & \underline{2.14\ldots} \\
2.5\,)5.81 & 2.6\,)5.80 & 2.7\,)5.79
\end{array}
$$

Comparing these three quotients, we conclude that the density is (2.2 $\pm$0.1) g/ml. In other words, the density, like the volume, is known to 2 significant figures.

This example illustrates a general rule for the multiplication or division of inexact numbers. In general, we should **retain in the product or quotient the number of significant figures present in the least precise of the numbers**. Applying this rule, we deduce that:

$(6.10 \times 10^3)\,(2.08 \times 10^{-4})$	has	3 significant figures in answer
5.92×3.0	has	2 significant figures in answer
$8.2/3.194$	has	2 significant figures in answer

The rule stated above can save us considerable time in carrying out multiplications or divisions. Suppose we are asked to divide 8.2 by 3.194. realizing that the answer can be given only to two significant figures, we might round off 3.194 to 3.2 before performing the division, thereby saving considerable time.*

Example 4.2 Carry out the following operations, retaining the correct number of significant figures:
  a. 6.19×2.8
  b. $3.18/1.702$
  c. $(4.10 \times 3.02 \times 10^9)/1.5$

Solution

  a. Since we are justified in retaining only two significant figures in our answer, we round off before multiplying and write:

$$6.2 \times 2.8 = 17$$

  b. $3.18/1.702 = 3.18/1.70 = 1.87$
  c. $(4.1 \times 3.0 \times 10^9)/1.5 = 8.2 \times 10^9$

---

*Many people prefer to retain one extra digit in carrying out a multiplication or division. In this case, they would round off 3.194 to 3.19 rather than 3.2, carry out the multiplication, and round off the answer to 2 significant figures. This procedure will change, at most, the last digit of the answer, perhaps by 2 or 3 units.

We can use the rules governing significant figures to guide us in designing laboratory experiments. If we are to carry out two or more measurements and combine them by multiplying or dividing to obtain a final result, the precision of that result will be governed by that of the least precise measurement. Suppose we are asked to determine the density of a liquid, using a triple beam balance (± 0.01 g) and a graduated cylinder (± 0.1 ml). The precision of the density will be limited by the comparatively large uncertainty in the volume measurement. If we need to know the density more exactly, we must resort to a more precise device for measuring volumes, such as a pipet or volumetric flask (± 0.01 ml). It would be fallacious to suppose that we could improve the precision by using an analytical balance capable of weighing to ± 0.0001 g and then measuring volume to ± 0.1 ml!

One final comment is in order concerning the rule that we have given for determining the number of significant figures in a product or a quotient. It is, at best, an approximation, albeit a convenient one. At times, it can lead to absurd situations. Let us suppose that two students, asked to determine the density of a liquid, obtain the following results:

	Mass	Volume	Density
Student 1	10.20 g	10.1 ml	1.01 g/ml
Student 2	10.10 g	9.9 ml	1.01 g/ml

Should the second student drop the last digit in his calculated density, reporting it as 1.0 g/ml, since there are only two significant figures in his volume? Common sense tells us that he should not. Like the other student, he has measured the volume to about ± 1 per cent. It is quite reasonable for him to report his answer to ± 1 per cent, i.e., as 1.01 g/ml rather than 1.0 g/ml, which would imply an uncertainty of ± 10 per cent.

Situations as extreme as this do not arise too frequently. They do, however, point out the limitations of using significant figures as a measure of precision, and suggest that a more exact treatment of experimental errors would be in order (Chapter 11).

EXERCISES

Carry out the following operations, giving answers to the correct number of significant figures.

1. $(2.49 \times 10^{-3})\ (3.81)$

2. 6.4023×19

3. 0.00481×212

4. $(3.18 \times 10^{-3})^2$

5. $7.17/6.2$

6. $8.73/5.198$

7. $\dfrac{6.48 \times 1.92}{5.2}$

8. $\dfrac{(8.10 \times 10^7)\ (4.43 \times 10^{-4})}{6.191 \times 10^2}$

4.3 ADDITION AND SUBTRACTION

When we add (or subtract) inexact numbers, we apply the general principle that the sum (or difference) *cannot have an absolute precision greater than that of the least precise number used in the computation.* To illustrate this point, suppose we add 1.32 g of sodium chloride and 0.006 g of potassium chloride to 28 g of water. How should we express the total mass of the resulting solution? The implied precisions of these masses are 0.01 g, 0.001 g, and 1 g, respectively.

sodium chloride:	1.32	± 0.01 g
potassium chloride:	0.006	± 0.001 g
water:	28	± 1 g
total mass:	29	± 1 g

The sum of the masses cannot be more precise than that of the water (± 1 g). We should write the total mass as 29 g, rather than 29.3 g, 29.326 g, or some other number.

You will note from this example that in adding (or subtracting) experimental quantities, it is *not* true that the number of significant figures in the answer is governed by the quantity having the fewest significant figures. The mass of potassium chloride is known to only 1 significant figure, while the sum can be expressed to 2 significant figures. It is the absolute precision (± 1 g of water) which determines the number of significant figures in the sum.

Example 4.3 Carry out the following operations, being careful to give answers to the correct number of significant figures:
   a. 6.82 g + 2.111 g + 1268 g
   b. 213 g − 0.01 g
   c. 5.19×10^{-2} cc + 1.83 cc + 2.19×10^{2} cc

Solution
   a. Since "1268 g" has the lowest absolute precision (± 1 g), the sum cannot have a greater precision. Therefore, we write:
$$7 \text{ g} + 2 \text{ g} + 1268 \text{ g} = 1277 \text{ g}$$
   b. The quantity 213 g is known only to ± 1 g, so subtracting 0.01 g from it has no effect. 213 g − 0 g = 213 g
   c. In order to obtain the sum, we first write out the numbers in decimal form:

$$5.19 \times 10^{-2} \text{ cc} = \quad 0.0519 \text{ cc} \pm 0.0001 \text{ cc}$$

$$1.83 \quad \text{cc} \pm 0.01 \text{ cc}$$

$$2.19 \times 10^{2} \text{ cc} = 219 \quad \text{cc} \pm 1 \text{ cc}$$

Since the sum cannot have a precision greater than 1 cc, we write:

$$0 + 2\,cc + 219\,cc = 221\,cc$$

EXERCISES

1. $9.10 + 6.231 = ?$
2. $8.162 - 2.39 = ?$
3. $4.30 + 29.1 + 0.345 = ?$
4. $6.23 + 915 - 12.7 = ?$

5. $2.02 \times 10^2 - 9.6 \times 10^1 = ?$
6. $3.18 \times 10^{-1} + 1.6 \times 10^{-2} = ?$
7. $(6.40 \times 12.1) - 2.19 = ?$
8. $\dfrac{3.18 \times 2.4}{1.92} + 0.17 = ?$

4.4   ROUNDING OFF NUMBERS

In calculations from experimental data, it is often necessary to drop one or more digits to obtain an answer with the appropriate number of significant figures. The rules which are followed in rounding off are:

1. If the first digit dropped is less than 5, leave the preceding digit unchanged.
2. If the first digit dropped is greater than 5, increase the preceding digit by 1.
3. If the first digit dropped is 5, leave the preceding digit unchanged if it is odd; increase it by 1 if it is even. (This means that, on the average, we will increase the retained digit half of the time and leave it unchanged half of the time.)

Example 4.4   Round off each of the following to 3 significant figures:

a. 6.167   d. 3135
b. 2.132   e. 6.19468
c. 0.002245

Solution

a. 6.17
b. 2.13
c. Here, since the digit to be retained, 4, is even, we increase it by 1 to give 0.00225.
d. The digit to be retained, 3, is odd; 3130.
e. 6.19 (the first digit dropped, 4, is less than 5).

EXERCISES

Round off 4.3154652 to the following numbers of significant figures:

1. 7
2. 6
3. 5
4. 4

5. 3
6. 2
7. 1

4.5   LOGARITHMS, ANTILOGARITHMS

Clearly, the precision of a number will govern the precision to be associated with its logarithm. This principle leads to the general rule that we should **retain in the mantissa of the logarithm the same number of significant figures as there are in the number itself.** Thus, we have:

$$\log 3.000 = 0.4771 \qquad \log 3.000 \times 10^5 = 5.4771$$

$$\log 3.00 \ = 0.477 \qquad \log 3.00 \times 10^3 \ = 3.477$$

$$\log 3.0 \ \ = 0.48 \qquad \log 3.0 \times 10^2 \ \ = 2.48$$

$$\log 3 \ \ \ = 0.5 \qquad \log 3 \times 10^{-4} \ \ = 0.5 - 4 = -3.5$$

Notice that, unless the characteristic is zero, there will be one more digit in the logarithm than in the number itself. This is quite reasonable, since the characteristic serves only to fix the position of the decimal point or the power of 10.

When we are asked to find the number corresponding to a certain logarithm, we follow a rule entirely analogous to that stated above. We **retain in the antilogarithm the same number of significant figures that we have in the mantissa of the logarithm.**

$$\text{antilog } 0.3010 = 2.000 \qquad \text{antilog } 5.3010 \qquad = 2.000 \times 10^5$$

$$\text{antilog } 0.301 \ = 2.00 \qquad \text{antilog } 1.301 \qquad = 2.00 \times 10^1$$

$$\text{antilog } 0.30 \ \ = 2.0 \qquad \text{antilog } 2.30 \qquad = 2.0 \times 10^2$$

$$\text{antilog } 0.3 \ \ \ = 2 \qquad \text{antilog } (0.3 - 4) = 2 \times 10^{-4}$$

Example 4.5   Find, using a 4-place log table:
   a.   log 6.19
   b.   log 1.3 $\times$ 10^{-4}
   c.   antilog 1.62

Solution

a. Referring to the table of logarithms in Appendix 2, we find:

$$\log 6.190 = 0.7917$$

Hence, $\log 6.19 = 0.792$ (3 significant figures)

b. $\log 1.300 = 0.1139$
   $\log 1.3 \quad = 0.11$
   $\log 1.3 \times 10^{-4} = 0.11 - 4 = -3.89$

c. Scanning down the column labeled "0," we find that:

$$\log 4.100 = 0.6128$$

$$\log 4.200 = 0.6232$$

We deduce that the 2-digit number whose logarithm is closest to 0.62 is 4.2:

$$\log 4.2 = 0.62$$

$$\text{antilog } 1.62 = 4.2 \times 10^1$$

EXERCISES

Evaluate to the correct number of significant figures:

1. $\log 1.602$
2. $\log 5.2$
3. $\log 2.18 \times 10^3$
4. $\log 4.9 \times 10^{-4}$
5. antilog 2.0
6. antilog 0.185
7. antilog 3.20
8. antilog (-1.902)

PROBLEMS

Observe the rules of significant figures in expressing your answers.

4.1 A student finds that a sample of liquid weighing 12.562 g occupies a volume of 8.04 cc. Calculate the density of the liquid.

4.2 A block of metal with dimensions 5.2 cm $\times$ 2.1 cm $\times$ 4.6 cm weighs 109.82 g. Calculate the density of the metal.

4.3 A student weighs an empty flask, adds some potassium nitrate to it, and reweighs. The two masses are 12.162 g and 12.498 g.
    a. What is the mass of potassium nitrate?
    b. What volume does the potassium nitrate occupy? (density
       = 2.107 g/ml)

4.4 Calculate the total mass of a solution prepared by adding 12.0 g of sodium chloride, 4.28 g of potassium nitrate, and 1.03 g of potassium chromate to 6.00×10^2 g of water.

4.5 In order to calibrate a pipet, a student weighs the water which drains from it after it has been filled to the mark. He obtains a mass of 9.9654 g. The density of water is 0.9970 g/ml. What is the volume of the pipet?

4.6 The gram equivalent weight of an element can be defined as the weight that combines with eight grams of oxygen. A student finds that when a metal oxide sample weighing 2.148 g is reduced with hydrogen, the pure metal remaining weighs 1.509 g.
    a. What is the weight of oxygen in the sample?
    b. What is the gram equivalent weight of the metal?

4.7 A sample of a certain compound weighing 2.040 g is found by analysis to contain 0.721 g of carbon and 0.050 g of hydrogen. It is known that the only other element present is iodine.
    a. What is the weight of iodine in the sample?
    b. What are the percentages by weight of the three elements
       in the compound?

4.8 When carbon burns in air, 3.667 g of carbon dioxide are formed for every gram of carbon. What mass of carbon dioxide is formed from samples of carbon weighing:
    a. 2.000 g
    b.   6.2 g

4.9 The molarity of a solution of sodium chloride can be expressed as:

$$M = \frac{\text{no. of grams NaCl}/58.44}{\text{no. of liters solution}}$$

A solution is prepared by dissolving 12.08 g of sodium chloride to give 9.00×10^2 ml. What is the molarity of this solution?

4.10 A sample of gas weighing 1.602 g occupies 224 ml at a pressure of 749 mm Hg and a temperature of 100.0°C. Calculate the gram molecular weight (M) of the gas using the equation:

$$M = \frac{gRT}{PV}$$

where g = mass in grams
  T = temperature in °K = °C + 273.16
  P = pressure in atm. = pressure in mm Hg/760.0
  V = volume in ml
  R = gas constant = 82.06 ml atm/mole °K

4.11 A student measures the pH of a certain solution, using instruments of successively greater precision. The values he obtains are as follows:
  a. 4
  b. 4.1
  c. 4.12
  d. 4.118
Using the definition: pH = −log (conc. H$^+$), calculate the concentration of H$^+$ corresponding to each measurement.

4.12 The equation relating the vapor pressure, P, of a liquid to the temperature can be written in the form:

$$\log P = \frac{-\Delta H}{2.303\, RT} + B$$

where R = 1.987 cal/°K, T = temperature in °K = °C + 273.16, ΔH = heat of vaporization in calories, and B is a constant known to be 5.169 for a particular liquid (when the vapor pressure is expressed in mm Hg).

If the vapor pressure of the liquid is 12.8 mm Hg at 25.00°C, what is the numerical value of ΔH?

4.13 The energy change, ΔE, in a nuclear reaction is related to the difference in mass, Δm, between products and reactants by the equation:

$$\Delta E = 2.15 \times 10^{10}\, \frac{kcal}{g} \times \Delta m$$

In a certain nuclear reaction, 225.9971 g of radium decomposes to give 221.9703 g of radon and 4.0015 g of helium. Calculate ΔE for this reaction.

UNIT CONVERSIONS

Many problems in general chemistry involve little more than a change in units. They are readily solved by the so-called "conversion factor method." While the mathematics of this method is extremely simple, beginning students often fail to appreciate its general applicability.

To illustrate the conversion factor approach, let us apply it to a particularly simple problem. Suppose we are asked to convert a length of 22 inches into feet. To do this, we make use of the conversion factor:

$$1 \text{ ft} = 12 \text{ in} \tag{5.1}$$

Dividing both sides of this equation by 12 in gives a quotient which is equal to unity:

$$\frac{1 \text{ ft}}{12 \text{ in}} = 1$$

If we now multiply 22 in by this quotient, we do not change the value of the length, but we do accomplish the desired conversion of units:

$$22 \text{ in} \times \frac{1 \text{ ft}}{12 \text{ in}} = 1.8 \text{ ft}$$

The conversion factor given by Equation 5.1 can be used equally well to convert a length given in feet, let us say 2.5 ft, to inches. In this case, we divide both sides of the equation by 1 ft to obtain:

$$\frac{12 \text{ in}}{1 \text{ ft}} = 1$$

Multiplying 2.5 ft by the ratio $\dfrac{12 \text{ in}}{1 \text{ ft}}$ converts the length from feet to inches:

$$2.5 \text{ ft} \times \frac{12 \text{ in}}{1 \text{ ft}} = 30 \text{ in}$$

Notice that a single conversion factor (e.g., 1 ft = 12 in) will always give us two quotients (1 ft/12 in or 12 in/1 ft) which are equal to unity. In making a conversion, we choose the quotient which will enable us to cancel out the unit that we wish to get rid of.

This approach is particularly valuable when we are dealing with unfamiliar units (Example 5.1).

Example 5.1   Convert 8.32×10^7 ergs to calories, using the conversion factor 1 erg = 2.39×10^{-8} cal.

Solution   To make effective use of the conversion factor approach, we need a quotient in which calories appear in the numerator and ergs in the denominator. This quotient is:

$$\frac{2.39 \times 10^{-8} \text{ cal}}{1 \text{ erg}} = 1$$

Multiplying 8.32×10^7 ergs by this quotient:

$$8.32 \times 10^7 \text{ ergs} \times \frac{2.39 \times 10^{-8} \text{ cal}}{1 \text{ erg}} = 1.99 \text{ cal}$$

Multistep conversions are readily accomplished by this method. To find the number of seconds in three days, using the conversion factors:

$$1 \text{ day} = 24 \text{ hrs}; \qquad 1 \text{ hr} = 60 \text{ min}; \qquad 1 \text{ min} = 60 \text{ sec}$$

we might first convert days to hours by multiplying by the quotient 24 hrs/1 day:

$$3 \text{ days} \times \frac{24 \text{ hrs}}{1 \text{ day}} = 72 \text{ hrs}$$

then convert to minutes, making use of the second conversion factor:

$$72 \text{ hrs} \times \frac{60 \text{ min}}{1 \text{ hr}} = 4320 \text{ min}$$

and, finally, change from minutes to seconds:

$$4320 \text{ min} \times \frac{60 \text{ sec}}{1 \text{ min}} = 2.59200 \times 10^5 \text{ sec}$$

In carrying out a multistep conversion, it is not, of course, necessary to solve for intermediate answers, as we did in the above example. Indeed, if successive multiplications and divisions are to be carried out with a slide rule, we often save time by setting up the problem on a single line, as in Examples 5.2 and 5.3.

Example 5.2   Given that:

1 atm = 760 mm Hg;   1 lit = 1000 ml;   1 lit atm = 24.22 cal

Convert 1.99 cal to the energy unit ml × mm Hg.

Solution   In problems of this type, it is helpful to analyze the successive conversions before putting any numbers on paper. The indicated steps are:
    (1) Convert 1.99 cal to liter atm   (1 lit atm = 24.22 cal)
    (2) Convert lit atm to ml atm   (1 lit = 1000 ml)
    (3) Convert ml atm to ml × mm Hg (1 atm = 760 mm Hg)

$$1.99 \text{ cal} \times \underset{(1)}{\frac{1 \text{ lit atm}}{24.22 \text{ cal}}} \times \underset{(2)}{\frac{1000 \text{ ml}}{1 \text{ lit}}} \times \underset{(3)}{\frac{760 \text{ mm Hg}}{1 \text{ atm}}}$$

$$= 6.24 \times 10^4 \text{ ml} \times \text{mm Hg}$$

Example 5.3   The density of mercury (Hg) is 13.6 g/ml. What is the mass in pounds of 1.50 lit of mercury?

Solution   We are, in effect, asked to convert liters of mercury to pounds. A few moments reflection might suggest the following three-step path:
    (1) lit Hg → ml Hg;   (1 lit = 1000 ml)

    (2) ml Hg → g Hg;   (1 ml Hg ≏ 13.6 g)

    (3) g Hg → lb Hg;   (1 lb = 453.6 g)

$$1.50 \text{ lit} \times \underset{(1)}{\frac{1000 \text{ ml}}{1 \text{ lit}}} \times \underset{(2)}{\frac{13.6 \text{ g}}{1 \text{ ml}}} \times \underset{(3)}{\frac{1 \text{ lb}}{453.6 \text{ g}}} = 45.0 \text{ lb}$$

In this problem, the symbol ⁐, which means "equivalent to," was used. Mathematically, conversion factors relating equivalent quantities are handled in the ordinary manner.

The conversion factor approach can be applied to a wide variety of problems in chemistry. As a simple example, consider how one might calculate the number of grams in 2.60 moles of $CaCl_2$, given that:

$$1 \text{ mole } CaCl_2 = 111 \text{ g}$$

Dividing both sides of this equation by 1 mole, we obtain:

$$\frac{111 \text{ g}}{1 \text{ mole}} = 1$$

Hence: no. g $= 2.60 \text{ moles} \times \dfrac{111 \text{ g}}{1 \text{ mole}} = 289 \text{ g}$

Example 5.4 illustrates the application of the conversion factor approach to the calculation of gram equivalent weights of elements.

Example 5.4 The gram equivalent weight of an element can be defined as the weight that combines with eight grams of oxygen. A student finds that a sample of a certain metal oxide weighing 1.468 g yields 1.214 g of metal when reduced by hydrogen. Calculate the gram equivalent weight of the metal.

Solution We are required to find the weight of metal which is chemically equivalent to eight grams of oxygen. We need a conversion factor relating grams of metal to grams of oxygen. From the data, we deduce that the metal oxide sample must have contained:

$$1.468 \text{ g} - 1.214 \text{ g} = 0.254 \text{ g oxygen}$$

Hence: 1.214 g metal ⁐ 0.254 g oxygen

$$\text{GEW metal} = 8.000 \text{ g oxygen} \times \frac{1.214 \text{ g metal}}{0.254 \text{ g oxygen}} = 38.2 \text{ g metals}$$

Conversion factors are particularly useful in solving problems dealing with chemical equations. The coefficients of a balanced equation give directly the conversion factors relating the numbers of moles of different substances participating in a reaction. The balanced equation:

$$2\,C_2H_6(g) + 7\,O_2(g) \rightarrow 4\,CO_2(g) + 6\,H_2O(l)$$

tells us that in this reaction, 2 moles of ethane, C_2H_6, 7 moles of oxygen, 4 moles of carbon dioxide, and 6 moles of water are chemically equivalent to each other.

2 moles C_2H_6 ⇌ 7 moles O_2 ⇌ 4 moles CO_2 ⇌ 6 moles H_2O

If we were asked to calculate the number of moles of C_2H_6 required to react with 16.0 moles of O_2, we would immediately write:

$$16.0 \text{ moles } O_2 \times \frac{2 \text{ moles } C_2H_6}{7 \text{ moles } O_2} = 4.57 \text{ moles } C_2H_6$$

Furthermore, since the mass in grams of one mole of any species is equal to its gram formula weight, we have:

1 mole C_2H_6 = 2(12.0 g) + 6(1.0 g) = 30.0 g C_2H_6

1 mole O_2 = 2(16.0 g) = 32.0 g O_2

1 mole CO_2 = 12.0 g + 2(16.0 g) = 44.0 g CO_2

1 mole H_2O = 2(1.0 g) + 16.0 g = 18.0 g H_2O

Using these conversion factors in combination with the mole relationships given by the balanced equation, we can work a variety of problems of the mole-mole, gram-mole, or gram-gram type (Example 5.5).

Example 5.5 Given the balanced equation for the combustion of ethane, calculate:
   a. The number of grams of O_2 required to react with 1.60 moles of C_2H_6.
   b. The number of moles of CO_2 formed from 12.0 g of O_2.
   c. The number of grams of C_2H_6 required to form 2.92 g of H_2O.

Solution
   a. We can regard this as a two-step conversion. We first convert 1.60 moles of C_2H_6 to moles of O_2, using the coeffi-

cients of the balanced equation (2 moles $C_2H_6 \backsimeq 7$ moles O_2). Then we convert to grams of O_2 (1 mole $O_2 = 32.0$ g O_2).

$$1.60 \text{ moles } C_2H_6 \times \frac{7 \text{ moles } O_2}{2 \text{ moles } C_2H_6} \times \frac{32.0 \text{ g } O_2}{1 \text{ mole } O_2} = 179 \text{ g } O_2$$

b. Proceeding in a manner entirely analogous to that in step (a), we convert

(1) g of $O_2 \rightarrow$ moles of O_2;   (1 mole $O_2 = 32.0$ g O_2)

(2) moles $O_2 \rightarrow$ moles of CO_2;   (7 moles $O_2 \backsimeq 4$ moles CO_2)

$$12.0 \text{ g } O_2 \times \frac{1 \text{ mole } O_2}{32.0 \text{ g } O_2} \times \frac{4 \text{ moles } CO_2}{7 \text{ moles } O_2} = 0.214 \text{ mole } CO_2$$

c. This problem may appear to be somewhat more complex, since it is not entirely obvious what path to follow. We might analyze the problem in either of two equivalent ways.

1. Realizing that the coefficients of the balanced equation give us a conversion factor relating *moles* of C_2H_6 to *moles* of H_2O, we could work through moles as an intermediate, following the path:

$$g\, H_2O \xrightarrow{\;a\;} \text{moles } H_2O \xrightarrow{\;b\;} \text{moles } C_2H_6 \xrightarrow{\;c\;} g\, C_2H_6$$

Hence: $2.92 \text{ g } H_2O \times \underset{(a)}{\frac{1 \text{ mole } H_2O}{18.0 \text{ g } H_2O}} \times \underset{(b)}{\frac{2 \text{ moles } C_2H_6}{6 \text{ moles } H_2O}}$

$$\times \underset{(c)}{\frac{30.0 \text{ g } C_2H_6}{1 \text{ mole } C_2H_6}} = 1.62 \text{ g } C_2H_6$$

2. We look for a conversion factor relating g of H_2O to g of C_2H_6, realizing that:

$$2 \text{ moles } C_2H_6 \backsimeq 6 \text{ moles } H_2O$$

Since one mole of C_2H_6 weighs 30.0 g and 1 mole of H_2O weighs 18.0 g, we have:

$$2(30.0 \text{ g}) \, C_2H_6 \backsimeq 6(18.0 \text{ g}) H_2O$$

$$60.0 \text{ g } C_2H_6 \backsimeq 108 \text{ g } H_2O$$

Hence: $2.92 \text{ g H}_2\text{O} \times \dfrac{60.0 \text{ g C}_2\text{H}_6}{108 \text{ g H}_2\text{O}} = 1.62 \text{ g C}_2\text{H}_6$

This approach can, of course, be used for any multistep conversion. It involves combining one or more conversion factors to find the particular factor that we need. In part (b), we could have reasoned that:

$$7 \text{ moles O}_2 \cong 4 \text{ moles CO}_2$$

but, since 1 mole of $O_2 = 32.0$ g of O_2:

$$7(32.0 \text{ g O}_2) \cong 4 \text{ moles CO}_2$$

$$224 \text{ g O}_2 \cong 4 \text{ moles CO}_2$$

Hence: $12.0 \text{ g O}_2 \times \dfrac{4 \text{ moles CO}_2}{224 \text{ g O}_2} = 0.214 \text{ mole CO}_2$

Conversion factors can also be used in working problems dealing with energy relationships in chemical reactions (Example 5.6).

Example 5.6 For the reaction:

$$C_3H_8(g) + 5 O_2(g) \rightarrow 3 CO_2(g) + 4 H_2O(l)$$

$\Delta H = -525$ kcal. In other words, 525 kcal of heat is evolved when one mole of C_3H_8 burns. Calculate:
   a. The amount of heat evolved when one gram of C_3H_8 burns.
   b. ΔH when one mole of CO_2 is formed.

Solution
   a. Our conversion factor is:

$$-525 \text{ kcal} \cong 1 \text{ mole C}_3\text{H}_8 = 44.0 \text{ g C}_3\text{H}_8$$

For one gram of C_3H_8, we have:

$$\Delta H = 1.00 \text{ g C}_3\text{H}_8 \times \dfrac{-525 \text{ kcal}}{44.0 \text{ g C}_3\text{H}_8} = -11.9 \text{ kcal}$$

i.e., 11.9 kcal of heat is evolved.

b. -525 kcal $\,\widehat{=}\,$ 1 mole C_3H_8

1 mole C_3H_8 $\,\widehat{=}\,$ 3 moles CO_2

$$\Delta H = 1 \text{ mole-CO}_2 \times \frac{1 \text{ mole-C}_3\text{H}_8}{3 \text{ moles-CO}_2} \times \frac{-525 \text{ kcal}}{1 \text{ mole-C}_3\text{H}_8}$$

$$= -175 \text{ kcal}$$

The principal advantage of the conversion factor approach is that it forces us to analyze a problem to determine what path we are going to follow to get from the information we are given to the quantity which is required. It has the additional advantage of setting up the arithmetic in a form which is ready-made for slide rule calculations.

PROBLEMS

Solve each of the following problems using the conversion factor approach. If the appropriate conversion factors are not given, look them up in your text or in a handbook.

5.1   A strip of tin foil has a length of 3.28 cm. Express its length in:
      a. mm   c. inches
      b. meters   d. Angstroms

5.2   The volume of a sodium atom is calculated to be 26.9 $(\text{Å})^3$. Calculate its volume in:
      a. cm^3 c. in^3
      b. liters

5.3   A certain gas exerts a pressure of 618 mm Hg. Express its pressure in:
      a. atm
      b. dynes/cm^2

5.4   The density of water is 62.5 lb/ft^3. Show that it has a density of 1.00 g/cm^3.

5.5   An oxygen molecule at 25°C has an average velocity of 4.82×10^4 cm/sec. What is its velocity in miles/hr?

5.6   The gram equivalent weight of a metal can be taken to be the weight that combines with 16.0 g of sulfur. It is found that when 1.200 g of a certain metal reacts with sulfur, 1.582 g of metal sulfide is formed. Calculate the gram equivalent weight of the metal.

5.7 The gram equivalent weight of an acid can be taken to be the weight that reacts with one mole of OH^-. It is found that 1.00 g of a certain acid reacts with 20.0 ml of a solution containing 0.200 mole/liter of OH^-. Calculate the gram equivalent weight of the acid.

5.8 The gram equivalent weight of a metal can be taken to be that weight which is plated from solution by one mole of electrons. It is found that 2.24×10^3 coulombs plate 0.518 g of a certain metal (1 mole electrons = 96,500 coulombs). What is the gram equivalent weight of the metal?

5.9 Calculate:
   a. The number of gram atomic weights in 12.8 g of copper (GAW Cu = 63.54 g).
   b. The number of atoms in 16.4 g of copper (1 GAW = 6.02×10^{23} atoms).
   c. The number of grams in 0.823 gram atomic weights of copper.

5.10 Calculate:
   a. The number of moles in 16.0 g of H_2O (1 mole H_2O = 18.0 g).
   b. The mass in grams of 8.46×10^9 molecules of H_2O (1 mole = 6.02×10^{23} molecules).
   c. The number of molecules in one gram of water.

5.11 Given the balanced equation:

$$2 \, AsH_3 \, (g) \rightarrow 2 \, As \, (s) + 3 \, H_2 \, (g)$$

in which the coefficients represent relative numbers of moles (1 mole AsH_3 = 77.9 g, 1 mole As = 74.9 g, 1 mole H_2 = 2.02 g), calculate:
   a. The number of moles of AsH_3 required to form 0.198 mole of As.
   b. The number of moles of H_2 produced from 1.29×10^{-4} mole of AsH_3.
   c. The number of grams of AsH_3 required to form 1.48 mole of H_2.
   d. The number of moles of As formed from 16.0 g of AsH_3.
   e. The number of grams of As produced simultaneously with 0.619 g of H_2.
   f. The number of grams of H_2 formed from 1.80 g of AsH_3.

5.12 For the reaction:

$$CH_4(g) + 2 O_2(g) \rightarrow CO_2(g) + 2 H_2O(1)$$

the enthalpy change, ΔH, is −213 kcal per mole of CH_4. Using a table of atomic weights where necessary, calculate:

      a. ΔH in kcal per mole of O_2.

      b. ΔH in kcal per gram of CH_4.

      c. ΔH in kcal when 6.12 g of H_2O is formed.

      d. ΔH in cal when 1.69 mole of CO_2 is formed.

      e. ΔH in BTU when 1.24×10^2 g of CH_4 burns (1 BTU = 252 cal).

5.13 Consider the reaction:

$$2 NH_3(g) \rightarrow N_2(g) + 3 H_2(g)$$

Suppose one starts with 4.08 moles of NH_3 and forms x moles of N_2. Express, in terms of x:

      a. The number of moles of NH_3 left.

      b. The number of moles of H_2 formed.

      c. The number of grams of N_2 formed.

5.14 In a certain nuclear reaction, the change in mass is 0.0210 g. Calculate the change in energy in:

      a. kcal

      b. ergs

      $(1 \text{ g} \triangleq 2.15 \times 10^{10} \text{ kcal})$

ALGEBRAIC EQUATIONS

Many problems in chemistry can be translated into algebraic equations. In this chapter, we shall consider methods of solution for those types of equations which are most frequently encountered in the first-year course. These include equations in which the highest power to which an unknown is raised is 1 (first degree equations) or 2 (second degree equations).

6.1 FIRST DEGREE EQUATIONS IN ONE UNKNOWN

Equations of this type are readily solved by using a fundamental principle of algebra which states that an equation remains valid if the same operation is performed on both sides. Several examples of first degree equations are listed below with their solutions.

	Equation	Operation
1.	$5x = 4$	Divide both sides by 5
	$x = 4/5 = 0.8$	
2.	$6 \times 10^{-4} = \dfrac{3 \times 10^{-1}}{x}$	Multiply both sides by x
	$(6 \times 10^{-4})x = 3 \times 10^{-1}$	Divide both sides by 6×10^{-4}
	$x = \dfrac{3 \times 10^{-1}}{6 \times 10^{-4}} = 0.5 \times 10^{3} = 5 \times 10^{2}$	

3. $3x = 4 - 5x$

   $8x = 4$ Add $5x$ to both sides

   $x = 4/8 = 0.5$ Divide both sides by 8

4. $\dfrac{5}{2x - 1} = \dfrac{2}{x + 4}$ Multiply both sides by

   $5(x + 4) = 2(2x - 1)$ $(2x - 1)(x + 4)$

   $5x + 20 = 4x - 2$

   $x + 20 = -2$ Subtract $4x$ from both sides

   $x = -22$ Subtract 20 from both sides

Examples 6.1 and 6.2 are typical of the type of general chemistry problems which lead to first degree equations in one unknown.

Example 6.1 The molarity, c, of a solution is defined by the equation:

$$c = n/V$$

where n is the number of moles of solute and V is the volume of the solution in liters. What volume of 6.0 molar solution can be prepared from 1.8 moles of solute?

Solution It is convenient to solve the equation for our unknown, V:

$$cV = n$$
$$V = n/c$$

Substituting numbers: $V = \dfrac{1.8}{6.0}\,\text{lit} = 0.30\,\text{lit}$

Example 6.2 The molarity, c, of a solution is related to its molality, m, by the expression:

$$c = \frac{md}{1 + mM_2/1000}$$

where d is the density in g/cc and M_2 is the molecular weight of solute. Calculate the molality of a 0.1000 molar solution which has a density of 1.004 g/cc and contains a solute of molecular weight 184.

Solution   Let us solve the equation for m. To do this, we first multiply both sides by the quantity $(1 + mM_2/1000)$:

$$c\,(1 + mM_2/1000) = md$$

$$c + \frac{cmM_2}{1000} = md$$

Subtracting the quantity $(cmM_2/1000)$ from both sides:

$$c = md - \frac{cmM_2}{1000}$$

$$= m\!\left(d - \frac{cM_2}{1000}\right)$$

Solving for m, we obtain:

$$m = \frac{c}{d - cM_2/1000}$$

Substituting numbers: $m = \dfrac{0.100}{1.004 - \dfrac{(0.100)(184)}{1000}} = \dfrac{0.100}{0.986} = 0.101$

EXERCISES

1. Solve for x:
   a. $4x = 6.0 \times 10^{-4}$
   b. $3x = \log 20$
   c. $1.98 = 2.64/x$
   d. $3x = 8x - 2.20 \times 10^3$
   e. $(2x - 4)/6 = (3 - 4x)/\log 10$
   f. $2/x - 1/6 = 1.0$

2. Solve each equation for the variable in bold type.

   a. $\dfrac{P_2}{P_1} = \dfrac{V_1}{\mathbf{V_2}}$

   b. $pV = nR\mathbf{T}$

   c. $\log P = \dfrac{-\mathbf{\Delta H}}{2.30\, RT} + B$

   d. $k = A\, e^{-\mathbf{\Delta E}/RT}$ (take logs of both sides)

6.2   SIMULTANEOUS FIRST DEGREE EQUATIONS IN TWO UNKNOWNS

Occasionally, problems in general chemistry lead naturally to a system of two equations in two unknowns. All the equations of this type that we will encounter will be of the first degree in both unknowns. A typical example would be:

$$2x + 5y = 15$$

$$3x - 4y = 1.8$$

A simple way to solve a pair of equations of this type is to use one of the equations to eliminate one variable. In the example just cited, we might solve the first equation for x:

$$2x = 15 - 5y; \qquad x = (15 - 5y)/2$$

Substituting in the second equation:

$$3 \frac{(15 - 5y)}{2} - 4y = 1.8$$

We now have a first degree equation in one unknown, which is readily solved by the method discussed in Section 6.1 to obtain:

$$y = 1.8$$

Having obtained a numerical value for y, we can easily find x, using either of the two original equations. For example, from the first equation:

$$2x + 5(1.8) = 15; \qquad 2x = 6; \qquad x = 3$$

Example 6.3 The atomic weight of chlorine is 35.453. It consists of two isotopes, chlorine-35 (A.W. = 34.969) and chlorine-37 (A.W. = 36.966). Calculate the fractions of the two isotopes.

Solution Let us choose the fraction of chlorine-35 to be x and that of chlorine-37 to be y. Realizing that these fractions must add up to unity, we have:

$$x + y = 1 \tag{1}$$

To obtain the other equation relating the variables, we note that the contribution of chlorine-35 to the atomic weight is 34.969 x, while that of chlorine-37 is 36.966 y.

$$34.969\, x + 36.966\, y = 35.453 \tag{2}$$

To solve this pair of equations, we eliminate x, using the first equation:

$$x = 1 - y$$

and substituting in the second equation:

$$34.969(1 - y) + 36.966\,y = 35.453 \qquad (3)$$

Solving for y:

$$1.997\,y = 0.484; \qquad y = 0.243$$

$$x = 1 - y = 0.757$$

You will note that we could have avoided introducing two variables if we had realized immediately that the fraction of chlorine-35 must be $1 - y$. This would have led directly to equation (3), which is a first degree equation in one unknown. Indeed, it is always possible to solve problems of this type in terms of a single variable if we are clever enough to see immediately a simple relationship between the two quantities involved.

EXERCISES

1. Solve for x and y:
   a. $3x - 9y = 16$

   $2x + 5y = -9$

   b. $\qquad\qquad x + y = 0.200$

   $0.0556x + 0.0153\,y = 0.00700$

2. Solve for x, y, and z:

   $$2x + 3y - z = 6$$
   $$x - 4y + 2z = 9$$
   $$3x + y - 3z = 4$$

(Note that it is always possible to solve independent simultaneous equations to obtain numerical answers for each variable, provided the number of equations equals the number of variables.)

6.3 SECOND DEGREE EQUATIONS IN ONE UNKNOWN

Equations of this type, often referred to as "quadratic equat arise frequently in problems dealing with chemical equilibrium. In cases, they can be solved by extracting the square root of both sides equation, as in the following examples:

Equation	Solution
1. $x^2 = 9.0 \times 10^{-4}$	1. $x = \pm 3 \times 10^{-2}$
2. $\dfrac{x^2}{(1-x)^2} = 4.0$	2. $\dfrac{x}{1-x} = \pm 2; \qquad x = \dfrac{2}{3}$ o

Quadratic equations, as in the above examples, always have two r When these equations arise in solving problems in general chemistr will turn out that one of the roots is physically absurd and can be carded (Example 6.4).

Example 6.4 The following relationship exists in any water solution saturated with barium sulfate:

$$(\text{conc. } Ba^{2+})(\text{conc. } SO_4^{2-}) = 1.5 \times 10^{-9}$$

One way to prepare such a solution is to dissolve barium sulfate in water, in which case the concentrations of Ba^{2+} and SO_4^{2-} must be equal.   Under these conditions, what is the concentration of Ba^{2+}?

Solution If we set the concentration of Ba^{2+} equal to x, then the concentration of SO_4^{2-} must also be x.

$$x^2 = 1.5 \times 10^{-9} = 15 \times 10^{-10}$$
$$x = \pm \sqrt{15} \times 10^{-5} = \pm 3.9 \times 10^{-5}$$

The answer -3.9×10^{-5}, although mathematically possible, is physically absurd. A species cannot have a concentration less than zero! We conclude that the concentration of Ba^{2+} must be 3.9×10^{-5} (moles/liter).

Many second degree equations which arise in solving general chemistry problems cannot be solved by the simple technique described above.

Consider, for example, the equation:

$$\frac{x^2}{1 - x} = 4$$

Clearly, we cannot solve for x by extracting the square root of both sides of the equation.

Any second degree equation in one unknown can be solved by applying the so-called "quadratic formula." To use this approach, we rewrite the equation, if necessary, to get it in the form:

$$ax^2 + bx + c = 0 \qquad (6.1)$$

where a, b, and c are numbers. The two roots of this equation are:

$$x = \frac{-b \pm \sqrt{b^2 - 4ac}}{2a} \qquad (6.2)$$

To illustrate how the quadratic formula is applied, let us use it to find the two values of x that satisfy the equation:

$$\frac{x^2}{1 - x} = 4$$

We first rewrite this equation to get it in the proper form:

$$x^2 = 4 - 4x$$
$$x^2 + 4x - 4 = 0$$

Comparing the equation just written to Equation 6.1, we deduce that:

$$a = 1; \qquad b = 4; \qquad c = -4$$

Consequently:

$$x = \frac{-4 \pm \sqrt{16 + 16}}{2} = \frac{-4 \pm \sqrt{32}}{2}$$

The square root of 32 is 5.66. So:

$$x = \frac{-4 \pm 5.66}{2} = \frac{1.66}{2} \text{ or } \frac{-9.66}{2}$$

$$x = 0.83 \text{ or } -4.83$$

Example 6.5 In a water solution of acetic acid, the following condition applies at equilibrium:

$$\frac{(\text{conc. H}^+)(\text{conc. Ac}^-)}{(\text{conc. HAc})} = 1.80 \times 10^{-5}$$

If we dissolve acetic acid in water, the reaction:

$$\text{HAc} \rightarrow \text{H}^+ + \text{Ac}^-$$

takes place until equilibrium is established. If we start with a concentration of HAc of 0.100 mole/liter, what will be the concentration of H$^+$ at equilibrium?

Solution If we allow x to be the concentration of H$^+$ at equilibrium, then that of Ac$^-$ must also be x and that of HAc, $0.100 - x$.

Original concentration		Change	Equilibrium concentration
HAc	0.100	$-x$	$0.100 - x$
H$^+$	0	$+x$	x
Ac$^-$	0	$+x$	x

Our equation then becomes:

$$\frac{x \cdot x}{0.100 - x} = 1.80 \times 10^{-5}$$

Rearranging to standard form:

$$x^2 = 1.80 \times 10^{-6} - 1.80 \times 10^{-5}x$$

$$x^2 + 1.80 \times 10^{-5}x - 1.80 \times 10^{-6} = 0$$

$$a = 1; \qquad b = 1.80 \times 10^{-5}; \qquad c = -1.80 \times 10^{-6}$$

Hence, $x = \dfrac{-1.80 \times 10^{-5} \pm \sqrt{(3.24 \times 10^{-10}) + 7.20 \times 10^{-6}}}{2}$

To obtain the square root, we note that: $3.24 \times 10^{-10} = 0.000324 \times 10^{-6}$. Hence:

$$3.24 \times 10^{-10} + 7.20 \times 10^{-6} = 0.000324 \times 10^{-6} + 7.20 \times 10^{-6}$$

$$= 7.20 \times 10^{-6}$$

$$x = \frac{-1.80 \times 10^{-5} \pm \sqrt{7.20 \times 10^{-6}}}{2}$$

$$= \frac{-1.80 \times 10^{-5} \pm 2.68 \times 10^{-3}}{2}$$

$$= \frac{-0.0180 \times 10^{-3} \pm 2.68 \times 10^{-3}}{2}$$

$$= \frac{2.66 \times 10^{-3}}{2} \text{ or } \frac{-2.70 \times 10^{-3}}{2}$$

Clearly, the second root is absurd, and we deduce that $x = 1.33 \times 10^{-3}$.

You will notice that the arithmetic here is tedious, to say the least. The quadratic formula itself is difficult enough to work with. To make matters worse, we have to add or subtract exponential numbers in two different cases. In Section 6.4, we will consider simpler methods of working problems of this type.

EXERCISES

1. Solve the following equations for x:

   a. $x^2 = 2.0 \times 10^{-4}$

   b. $3x^2 = 4.5 \times 10^{-9}$

   c. $x^2/(1 - x)^2 = 2.0$

   d. $\dfrac{(2x)^2}{(1 - x)^2} = 2.0 \times 10^{-3}$

   e. $x^2/(1 - x) = 0.30$

   f. $4x^2/(1 - 3x) = 1.8 \times 10^{-4}$

   g. $x^2/(0.10 - x) = 4.0 \times 10^{-7}$

   h. $\dfrac{x^4}{(2 - x)^2} = 1.0 \times 10^{-4}$

2. Solve Example 6.5 for an acid which has a concentration quotient (known as the ionization constant) of:

   a. 1.80×10^{-2}

   b. 1.80×10^{-8}

6.4 SOLUTION OF SECOND DEGREE EQUATIONS BY APPROXIMATION METHODS

Although it is always possible to solve a second degree equation with the aid of the quadratic formula, it is seldom convenient to do so. Example 6.5 illustrates the rather complicated and tedious arithmetic that is involved. Since second degree equations similar to that encountered in Example 6.5

arise very frequently in general chemistry, it is highly desirable to find simpler ways of solving them. One such approach, often referred to as the method of *successive approximations*, can be used with a wide variety of problems dealing with equilibria in solutions of weak electrolytes.

To illustrate how this method works, let us apply it to Example 6.5, where we were required to solve the equation:

$$\frac{x^2}{0.100 - x} = 1.80 \times 10^{-5}$$

Noting that the ionization constant of acetic acid, 1.80×10^{-5}, is a very small number, it seems reasonable to suppose that x, the concentration of H^+ produced when acetic acid ionizes, will be very small. In particular, it seems likely that x *will be very much smaller than 0.100*, the original concentration of acetic acid. If this is true, we would be justified in ignoring the x in the denominator of the above equation, writing:

$$\frac{x^2}{0.100} = 1.80 \times 10^{-5}$$

This approximate equation is readily solved for x:

$$x^2 = 1.80 \times 10^{-6}; \qquad x = \sqrt{1.80} \times 10^{-3} = 1.34 \times 10^{-3}$$

Let us compare this value of x, obtained by making the approximation $0.100 - x \approx 0.100$, to that obtained in Example 6.5, where we used the quadratic formula. The two numbers, 1.34×10^{-3} and 1.33×10^{-3} differ from each other by 1 part in 133, or less than 1 per cent. Errors of this order of magnitude are ordinarily acceptable in working problems dealing with weak electrolytes' equilibria, since the equilibrium constants themselves are seldom valid to better than ± 5 per cent.

Clearly, the simplifying approximation described above is valid in the particular case dealt with in Example 6.5. The question arises as to the general validity of this approach. Specifically, if we are dealing with an equation of the type:

$$\frac{x^2}{a - x} = K \tag{6.3}$$

we wish to know how large an error will be introduced by making the approximation:

$$a - x \approx a$$

and solving the resultant equation:

$$x^2 = aK \tag{6.4}$$

It can be shown (see Exercise 4) that, provided K is small compared to a, as is usually the case in problems we deal with in general chemistry:

$$\text{percent error} \approx 50 \left(\frac{K}{a}\right)^{1/2} \qquad (6.5)$$

Thus, if our equation is:

$$\frac{x^2}{0.10 - x} = 1.0 \times 10^{-5}$$

the percent error associated with the approximation will be:

$$50 \times \left(\frac{10^{-5}}{0.10}\right)^{1/2} = 50 \times (10^{-4})^{1/2} = 0.50 \text{ per cent}$$

while if the equation is:

$$\frac{x^2}{1.0 - x} = 1.0 \times 10^{-2}$$

$$\text{percent error} = 50 \times \left(\frac{1.0 \times 10^{-2}}{1.0}\right)^{1/2} = 5.0 \text{ per cent}$$

In most problems dealing with weak electrolyte equilibria, we will find the approximation described above to be valid. Occasionally, however, the percent error that arises from the approximation, as calculated by Equation 6.5, will exceed the limit of accuracy that we are allowed. If we require a more exact answer, we can always refine our calculations by making a second approximation, more nearly valid than the first. What we do here is to substitute for x, in the denominator of Equation 6.3, the value obtained by the first approximation. Solving the resultant equation for x will give us a number considerably closer to the true value. The technique is illustrated by Example 6.6.

Example 6.6 Solve the equation:

$$\frac{x^2}{1.0 - x} = 1.0 \times 10^{-2}$$

to obtain x accurate to ± 1 per cent.

Solution Our first approximation is:

$$1.0 - x \approx 1.0$$

The resultant equation is: $\quad x^2 = 1.0 \times 10^{-2}$

$$x = 1.0 \times 10^{-1}$$

But, as we saw from the discussion following Equation 6.5, this value is in error by about 5 per cent. To obtain a more accurate value, we substitute $x = 1.0 \times 10^{-1} = 0.10$ in our original equation and obtain:

$$\frac{x^2}{1.0 - 0.1} = 1.0 \times 10^{-2}$$

$$x^2 = 0.90 \times 10^{-2}$$

$$x = 0.95 \times 10^{-1} = 9.5 \times 10^{-2}$$

Note that the second approximation has changed our answer by 5 parts in 100, or about 5 per cent. We estimated that our first answer was in error by about 5 per cent, so it would appear that we are now very close to the true value of x.

If necessary, we could of course repeat this process of successive approximations to obtain a still more accurate answer. In practice, it is almost never necessary to do this, since the error decreases exponentially with the number of approximations. That is:

$$f_n = (f_1)^n \tag{6.6}$$

where f_1 is the fractional error (percent error/100) that we introduce by making the first approximation ($a - x \approx a$) and f_n is the fractional error remaining after n approximations. This means, for example, that even if the first approximation gave an error of 10 per cent ($f_1 = 0.10$), a second approximation would reduce the error to:

$$(0.10)^2 = 0.01$$

or 1 per cent.

The approximation method which we have described is by no means restricted to second degree equations. Indeed, it is particularly useful for higher degree equations where exact solutions are either very difficult or impossible to obtain. Consider, for example, the equation:

$$\frac{x^3}{(2.00 - x)^2(1.00 - x)} = 1.0 \times 10^{-3}$$

If we assume for the time being that x is small, relative to 2 or 1, we can write:

$$\frac{x^3}{2^2(1)} \approx 1.0 \times 10^{-3}$$

$$x^3 \approx 4.0 \times 10^{-3}$$

$$x \approx (4.0)^{1/3} \times 10^{-1} = 1.6 \times 10^{-1} = 0.16 \text{ (1st approximation)}$$

Substituting this approximate value of x in the original equation, we have:

$$\frac{x^3}{(2.00 - 0.16)^2(1.00 - 0.16)} = 1.0 \times 10^{-3}$$

$$x^3 = (1.84)^2(0.84) \times 10^{-3}$$

Solving:

$$x = 1.4 \times 10^{-1} = 0.14 \qquad \text{(2nd approximation)}$$

Finally, we should point out that certain equations of the second degree or higher can be solved by making approximations of a quite different type than the one we have discussed. Example 6.7 illustrates such a case.

Example 6.7 For the reaction:

$$H^+ + Ac^- \rightarrow HAc$$

the equilibrium constant is 5.6×10^4. That is, at equilibrium:

$$\frac{(\text{conc. HAc})}{(\text{conc. } H^+)(\text{conc. } Ac^-)} = 5.6 \times 10^4$$

If one starts with 1.0 M solutions of H^+ and Ac^-, what are the equilibrium concentrations of all three species?

Solution   If we let x = equilibrium concentration of HAc

$$\text{then equil. conc. } H^+ = 1.0 - x$$

$$\text{equil. conc. } Ac^- = 1.0 - x$$

Consequently, we have: $\dfrac{x}{(1.0 - x)^2} = 5.6 \times 10^4$

Now, since the equilibrium constant is very large, most of the H^+ and Ac^- must react to form HAc; *this means that $(1.0 - x)$ will be very small or that $x \approx 1$.* If we make this substitution in the numerator of the above equation:

$$\frac{1}{(1.0 - x)^2} = 5.6 \times 10^4$$

Solving: $(1 - x)^2 = \dfrac{1}{5.6 \times 10^4} = 1.8 \times 10^{-5} = 18 \times 10^{-6}$

$$1 - x = 4.2 \times 10^{-3} = \text{conc. Ac}^- = \text{conc. H}^+$$

$$\text{conc. HAc} = x = 1 - 4.2 \times 10^{-3} \approx 1$$

EXERCISES

1. Solve the following equations by making the approximation: $a - x \approx a$.

   a. $\dfrac{x^2}{1 - x} = 1.0 \times 10^{-6}$

   b. $\dfrac{x^2}{0.10 - x} = 1.0 \times 10^{-3}$

   c. $\dfrac{x^2}{0.20 - x} = 2.0 \times 10^{-4}$

2. Using Equation 6.5, estimate the errors involved in each part of Exercise 1. If a second approximation were made in each case, what would the error be?

3. Solve, to two significant figures:

   a. $\dfrac{x^3}{(2.0 - x)^2(1.0 + x)} = 2.0 \times 10^{-6}$

   b. $\dfrac{x^2}{1.00 - x} = 50 \quad (0 < x < 1)$

4. Derive Equation 6.5. To do this, first solve Equation 6.3 for x by the use of the quadratic formula and show that if $K \ll a$, $x = (aK)^{1/2} - K/2$. By how much does this answer differ from that obtained by solving Equation 6.4? Finally, note that:

$$\text{percent error in } x = \frac{100(\text{error in } x)}{x}$$

PROBLEMS

6.1   In a water solution in equilibrium with $PbCl_2$, the concentrations of Pb^{2+} and Cl^- are related by the expression:

$$(\text{conc. } Pb^{2+})(\text{conc. } Cl^-)^2 = 1.7 \times 10^{-5}$$

a. If the concentration of Cl^- is 1.0×10^{-2}, what is the concentration of Pb^{2+} ?

b. What is the concentration of Cl^- when that of Pb^{2+} is 1.0×10^{-1}?

c. What is the concentration of Pb^{2+} in a solution in which: $(\text{conc. } Cl^-) = 2 (\text{conc. } Pb^{2+})$?

6.2   The freezing point lowering of a water solution is given by the equation:

$$\text{f.p.l.} = \frac{(1.86°C)(\text{no. of g solute})}{M \text{ (no of kg water)}}$$

where M is the molecular weight.   Calculate M if a solution containing 10.0 g of solute in 90.0 g of water freezes at $-0.372°C$.

6.3   The mole fraction, X_2, of a solute in water solution is related to its molality, m, by the equation:

$$X_2 = \frac{m}{m + 1000/18.0}$$

Calculate the molality of a solute when the mole fraction is 0.10.

6.4   For the reaction: $2HI(g) \longrightarrow I_2(g) + H_2(g)$
the following relation holds at equilibrium:

$$\frac{(\text{conc. } H_2) \times (\text{conc. } I_2)}{(\text{conc. } HI)^2} = 0.20$$

If one starts with a concentration of HI of 0.50 mole/liter, what is the equilibrium concentration of H_2? (Note that for every mole of H_2 formed, one mole of I_2 is formed and two moles of HI are consumed.)

6.5 In a water solution saturated with both NiS and H_2S, the following relationships exist:

$$(\text{conc. } Ni^{2+})(\text{conc. } S^{2-}) = 1 \times 10^{-22}$$

$$(\text{conc. } H^+)^2 (\text{conc. } S^{2-}) = 1 \times 10^{-23}$$

Calculate the concentration of Ni^{2+} in a solution in which the concentration of H^+ is 1×10^{-9} M (mole/liter).

6.6 For a water solution of acetic acid:

$$\frac{(\text{conc. } H^+) \times (\text{conc. } Ac^-)}{(\text{conc. } HAc)} = 1.80 \times 10^{-5}$$

Calculate the concentration of H^+ when the concentration of Ac^- is five times that of HAc.

6.7 For the reaction: $HF \rightarrow H^+ + F^-$, at equilibrium:

$$\frac{(\text{conc. } H^+) \times (\text{conc. } F^-)}{(\text{conc. } HF)} = 7.0 \times 10^{-4}$$

If the original concentration of HF is 1.0 M:

    a. Calculate the equilibrium concentration of H^+, making the approximation $1.0 - x \approx 1.0$.
    b. Estimate the error involved in making this approximation.
    c. Calculate the equilibrium concentration of H^+, using the quadratic formula.

6.8 For the reaction: $NH_3 + H_2O \rightarrow NH_4^+ + OH^-$

at equilibrium: $\dfrac{(\text{conc. } OH^-) \times (\text{conc. } NH_4^+)}{(\text{conc. } NH_3)} = 1.80 \times 10^{-5}$

If the original concentration of NH_3 is 0.0100 M:

    a. Calculate the equilibrium concentration of NH_4^+, neglecting the x in the denominator.
    b. Use the value of concentration NH_4^+ calculated in part (a) to carry out a second approximation to obtain a more accurate value.
    c. Estimate the percent error in the values calculated in parts (a) and (b).

6.9 For the reaction: $H^+ + NH_3 \rightarrow NH_4^+$

at equilibrium: $\dfrac{(\text{conc. } NH_4^+)}{(\text{conc. } H^+) \times (\text{conc. } NH_3)} = 1.8 \times 10^9$

If one starts with a solution in which the concentrations of both H^+ and NH_3 are 1.0 M, calculate the concentration of H^+ at equilibrium.

6.10 For the reaction: $2SO_2(g) + O_2(g) \rightarrow 2SO_3(g)$

at equilibrium: $\dfrac{(\text{conc. } SO_3)^2}{(\text{conc. } SO_2)^2 \times (\text{conc. } O_2)} = 1.0 \times 10^{-4}$

If the original concentrations of SO_2 and O_2 are 2.0 and 1.0 mole/liter respectively, what is the concentration of SO_3 at equilibrium? Note that for every mole of SO_3 formed, one mole of SO_2 and $\frac{1}{2}$ mole of O_2 are consumed.

CHAPTER 7

FUNCTIONAL RELATIONSHIPS

A variable (y) is said to be a function of another variable (x) if, for various values of x, it is possible to establish corresponding values of y. In mathematical symbolism, we describe this situation by writing:

$$y = f(x)$$

The variable (x) to which we first assign numerical values is referred to as the **independent variable;** the other variable (y) is called the **dependent variable.**

Table 7.1 lists a few relatively simple functional relationships. For each function, we give the values of y calculated by substituting $x = 0$, 1, 2, and 3 in the corresponding equation.

Table 7.1	Examples of Functional Relationships			
1. $y = 3x$	if $x =$	0	1	2	3
	then $y =$	0	3	6	9
2. $y = 8x - 3$	if $x =$	0	1	2	3
	then $y =$	-3	5	13	21
3. $y = 4/x$	if $x =$	0	1	2	3
	then $y =$	—	4	2	4/3
4. $y = \log x$	if $x =$	0	1	2	3
	then $y =$	—	0	0.3010	0.4771

Equations relating one variable to another usually include one or more numbers. In the function $y = 3x$, the number 3 appears; in the function $y = 8x - 3$, we have two numbers, 8 and -3. These numbers are often referred to as **constants,** since they are the same for all values of x and y.

7.1 FUNCTIONAL RELATIONSHIPS IN GENERAL CHEMISTRY

Perhaps the simplest type of functional relationship is that in which y is **directly proportional** to x. The first relation listed in Table 7.1 ($y = 3x$) is an example of a direct proportionality. The general equation is:

$$y = ax \tag{7.1}$$

where a is a constant. An example of a direct proportionality that we encounter in general chemistry is that relating the rate of decomposition of the gaseous compound, N_2O_5, to its concentration:

$$\text{rate} = k \, (\text{conc. } N_2O_5)$$

The quantity k appearing in this equation is called the rate constant. At a given temperature, it has a fixed value which could be found by measuring the rate of reaction at one particular concentration of N_2O_5, i.e.:

$$k = \text{rate}/(\text{conc. } N_2O_5)$$

Another type of functional relationship which arises frequently in general chemistry is an **inverse proportionality.** The third relation listed in Table 7.1 ($y = 4/x$) is of this type. In general, y is inversely proportional to x if:

$$y = \frac{a}{x} \tag{7.2}$$

where a, once again, is a constant. The concentrations of H^+ and OH^- in aqueous solution at 25°C are inversely related to each other. In this case, the constant, often referred to as the ionization constant of water, has the numerical value 1.0×10^{-14}.

$$(\text{conc. } H^+) = \frac{1.0 \times 10^{-14}}{(\text{conc. } OH^-)}$$

The functional relationship which we will consider most extensively in this chapter is the **linear function,** which has the general form:

$$y = ax + b \tag{7.3}$$

where a and b are both constants. The phrase "linear function" is used because a straight line is obtained when y is plotted against x (see Section 7.2). The relationship:

$$y = ax$$

is, of course, a special case of a linear function, with $b = 0$.

One of the fundamental equations of thermodynamics, the so-called Gibbs-Helmholtz equation, can be considered as a linear function relating the free energy change, ΔG, to the absolute temperature, T:

$$\Delta G = \Delta H - T \, \Delta S \qquad (7.4)$$

The quantities ΔH and ΔS appearing in this equation are "constants" in the sense that they do not vary, at least to a first approximation, with temperature. These quantities are referred to as the enthalpy change and entropy change, respectively.

Example 7.1   Using Equation 7.4, determine:
   a. ΔG at 400°K for a reaction for which $\Delta H = 61.0$ kcal and $\Delta S = 0.020$ kcal/°K.
   b. ΔG at 500°K for a reaction for which $\Delta H = -32.0$ kcal and $\Delta G = -20.0$ kcal at 300°K.

Solution
   a. $\Delta G = 61.0$ kcal $- 400°K(0.020$ kcal/°K$) = 53.0$ kcal.
   b. One way to solve this problem is to first calculate ΔS by applying Equation 7.4 at 300°K, and then use this value to calculate ΔG at 500°K.
   At 300°K: -20.0 kcal $= -32.0$ kcal $- 300°K(\Delta S)$
   $\Delta S = -12.0$ kcal/300°K $= -0.040$ kcal/°K
   At 500°K: $\Delta G = -32.0$ kcal $- 500°K(-0.040$ kcal/°K$)$
   $= -12.0$ kcal

Another type of functional relationship which turns up repeatedly in general chemistry has the form:

$$\log y = \frac{-A}{x} + B \qquad (7.5)$$

An example is the relationship between the vapor pressure of a liquid (P) and the absolute temperature (T):

$$\log P = \frac{-\Delta H_{vap}}{2.30\, RT} + B \tag{7.6}$$

where ΔH_{vap} is the molar heat of vaporization (cal/mole) and R is the gas constant (1.99 cal/mole °K). The functional relationship between P and T is often expressed in terms of "final" and "initial" states. If we apply Equation 7.6 at two different temperatures, T_2 and T_1:

$$\log P_2 = \frac{-\Delta H_{vap}}{2.30\, RT_2} + B$$

$$\log P_1 = \frac{-\Delta H_{vap}}{2.30\, RT_1} + B$$

Subtracting: $\log P_2 - \log P_1 = \dfrac{\Delta H_{vap}}{2.30\, RT_1} - \dfrac{\Delta H_{vap}}{2.30\, RT_2} = \dfrac{\Delta H_{vap}}{2.3\, R}\left[\dfrac{1}{T_1} - \dfrac{1}{T_2}\right]$

$$\log \frac{P_2}{P_1} = \frac{\Delta H_{vap}}{2.3\, R}\left[\frac{T_2 - T_1}{T_2 T_1}\right] \tag{7.7}$$

Example 7.2 The vapor pressure of water at 298°K is 23.6 mm Hg; its heat of vaporization is 10,500 cal/mole. Calculate:

   a. The vapor pressure of water at 310°K.
   b. The temperature at which water has a vapor pressure of 30.0 mm Hg.

Solution

   a. $\log \dfrac{P_2}{P_1} = \dfrac{10,500}{(2.30)(1.99)}\left[\dfrac{310 - 298}{(310)(298)}\right]$

$$= \frac{10,500 \times 12}{2.30 \times 1.99 \times 310 \times 298} = 0.298$$

$$\log \frac{P_2}{23.6} = 0.298$$

Taking antilogs: $\dfrac{P_2}{23.6} = 1.99;$ $P_2 = 47.0$ mm Hg

b. $\log \dfrac{30.0}{23.6} = \dfrac{10,500}{(2.30)(1.99)} \dfrac{(T_2 - 298)}{298\, T_2}$

$$1.4771 - 1.3729 = 7.70 \dfrac{(T_2 - 298)}{T_2}$$

Solving: $T_2 = 302°K$

Certain of the quantities that we deal with in general chemistry are a function of more than one variable. A familiar example is the volume, V, of an ideal gas, which is a function of the number of moles, n, the temperature, T, and the pressure, P.

$$V = \dfrac{n\,R\,T}{P} \qquad (R = \text{gas law constant} = 0.0821 \text{ lit atm/mole°K}) \qquad (7.8)$$

If we specify the numerical values of two of the variables, it is possible to obtain a relationship between the other two variables. For example, if we are dealing with one mole of an ideal gas at one atmosphere pressure:

$$V = \dfrac{1 \text{ mole } (0.0821 \text{ lit atm/mole °K})\, T}{1 \text{ atm}} = 0.0821 \dfrac{\text{lit}}{°K}\, T$$

Again, for two moles of gas at 300°K, the relationship between volume and pressure is:

$$V = \dfrac{(2 \text{ moles}) (0.0821 \text{ lit atm/mole °K}) (300°K)}{P} = \dfrac{49.3 \text{ lit atm}}{P}$$

Throughout this section, we have referred to various "constants" which are used to express functional relationships. Frequently, these constants have units, as is the case with the gas law constant used in Equations 7.6 and 7.8:

$$R = \dfrac{1.99 \text{ cal}}{\text{mole °K}} = \dfrac{0.0821 \text{ lit atm}}{\text{mole °K}}$$

Here, we must be careful to express R in units which are consistent with those used in other terms appearing in the equation. In Equation 7.6, where ΔH is given in *cal/mole*, R should be expressed as *1.99 cal/mole*. In Equation 7.8, if V is given in *liters* and P in *atmospheres*, we should use $R = 0.0821$ *lit atm/mole °K*.

Certain of the "constants" that we work with in general chemistry are themselves functions of other quantities which do not appear in the equation. An example is the rate constant, k, for the decomposition of N_2O_5:

$$\text{rate of decomposition of } N_2O_5 = k \text{ (conc. } N_2O_5)$$

The numerical value of k depends upon the temperature at which the rate is measured. At 400°K, k is 1.3/sec; at 500°K, k is 170/sec. If we write the explicit equation:

$$\text{rate} = \frac{170 \text{ (conc. } N_2O_5)}{\text{sec}}$$

we must realize that this particular functional relationship is limited to one temperature, 500°K.

EXERCISES

1. For a certain reaction, ΔH is +23.0 kcal and ΔG is +17.0 kcal at 300°K. Using Equation 7.4, calculate:
    a. ΔS
    b. ΔG at 500°K
    c. T at which $\Delta G = 0$

2. The equilibrium constant for a reaction, K, varies with temperature, T, according to the relation:

$$\log \frac{K_2}{K_1} = \frac{\Delta H}{2.3R} \left(\frac{T_2 - T_1}{T_2 T_1} \right)$$

where ΔH is in the enthalpy change in cal/mole, R is the gas law constant (1.99 cal/mole). For a certain reaction, K is 1.0 at 300°K. Calculate K at 400°K if ΔH is:
    a. 0
    b. $+1.00 \times 10^4$ cal/mole
    c. -1.00×10^4 cal/mole

3. At 273°K and one atmosphere pressure, one mole of an ideal gas occupies 22.4 liters.
    a. Using Equation 7.8, evaluate R in lit atm/mole °K.
    b. Given that 1 lit atm = 24.2 cal, calculate R in cal/mole °K.

7.2 GRAPHS OF FUNCTIONS

Functional relationships between two variables can always be represented graphically. Ordinarily, we use the vertical axis for the dependent variable, y, and the horizontal axis for the independent variable, x. If we wish to graph the function:

$$y = 2.5\,x + 3.0$$

in the region between $x = 0$ and $x = 5$, we might first calculate y for $x = 0, 1, 2, 3, 4,$ and 5:

y	3.0	5.5	8.0	10.5	13.0	15.5
x	0	1	2	3	4	5

and then locate each of these points on the graph. The first point ($x = 0$, $y = 3.0$) is located on the y axis ($x = 0$), three units above the origin ($y = 3.0$). The second point ($x = 1$, $y = 5.5$) is located by moving out one unit horizontally from the origin ($x = 1$) and then moving 5.5 units vertically ($y = 5.5$). The graph which results when the six points located in this manner are connected is shown in Figure 7.1.

We note that the plot of the function:

$$y = 2.5\,x + 3.0$$

is a straight line which cuts the y axis at 3.0. The slope of this line may be found by dividing the difference between final and initial values of y by the difference between the corresponding x values.

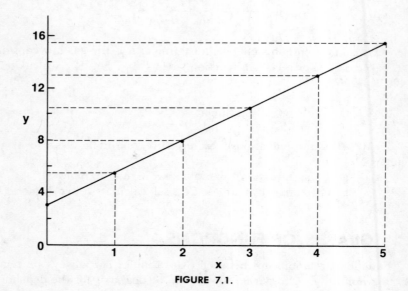

FIGURE 7.1.

$$\text{slope} = \frac{\Delta y}{\Delta x} = \frac{15.5 - 3.0}{5.0 - 0.0} = 2.5$$

We conclude that for this function, the **slope** is 2.5 and the **y intercept** (the value of y when $x = 0$) is 3.0. In general, we can say that the graph of the equation:

$$y = ax + b$$

is a straight line with a slope of a and an intercept of b. To show that this is the case, we first let $x = 0$, obtaining:

$$y_0 = a(0) + b; \qquad y_0 = b$$

when $x = 1$: $\qquad y_1 = a + b$

Subtracting: $\quad$ slope $= \dfrac{\Delta y}{\Delta x} = \dfrac{y_1 - y_0}{1 - 0} = \dfrac{(a + b) - b}{1 - 0} = a$

Many of the functions that we work with in general chemistry give plots which are smooth curves rather than straight lines. Examples include Equations 7.2 ($y = a/x$) and 7.5 ($\log y = -A/x + B$), which are plotted in Figure 7.2.

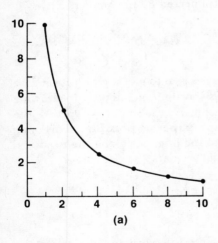

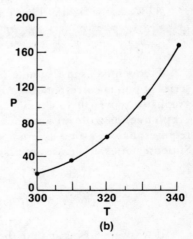

FIGURE 7.2. (a) $y = a/x$; ($a = 10$)

(b) $\log P = -\dfrac{A}{T} + B$

($A = 2400$, $B = 9.30$)

Example 7.3   Construct a graph of the function:

$$\log P = \frac{-2400}{T} + 9.30$$

in the region between $T = 300$ and $T = 340$.

Solution   In order to construct the graph, it will be necessary to locate several points within this range. To do this, we might calculate values of P at 10° intervals between 300 and 340°K:

$T = 300;$ $\log P = \dfrac{-2400}{300} + 9.30 = 1.30;$ $P = 20;$ (2 sig. fig.)

$T = 310;$ $\log P = \dfrac{-2400}{310} + 9.30 = 1.56;$ $P = 36$

$T = 320;$ $\log P = \dfrac{-2400}{320} + 9.30 = 1.80;$ $P = 63$

$T = 330;$ $\log P = \dfrac{-2400}{330} + 9.30 = 2.03;$ $P = 110$

$T = 340;$ $\log P = \dfrac{-2400}{340} + 9.30 = 2.24;$ $P = 170$

These points are plotted and connected by the smooth curve shown at the right of Figure 7.2.

Frequently, as in Example 7.3, we are asked to plot a graph through a series of points corresponding to successive values of x and y. In drawing graphs to represent functional relationships, we attempt to **spread the graph over as wide an area of the graph paper as possible.** This often requires that we assign different values to the intervals along the two axes. Suppose, for example, we are asked to plot the data:

y	0	10	20	30	40	50
x	0	1	2	3	4	5

If we were to insist that the same interval represent one unit along both axes, we would get the straight line shown in Fig. 7.3a. This graph would be rather difficult to read, since it nearly coincides with the y axis. It would be better to let an interval along the y axis represent 10 units, as

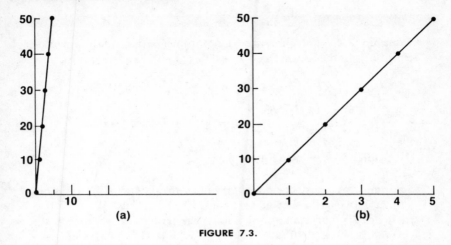

FIGURE 7.3.

compared to 1 unit along the *x* axis, thereby obtaining the straight line shown in Figure 7.3b.

Following this same principle, it is often inadvisable to let the intersection of the two axes represent the point (0,0). Suppose we were asked to plot the data:

y	200	205	210	215	220
x	100	101	102	103	104

It would be unreasonable, to say the least, to start our graph at $y = 0$, $x = 0$. If we did, we would get the nearly invisible line shown in Figure 7.4a. The space available to us could be utilized much more effectively by starting at the lower left with $y = 200$, $x = 100$.

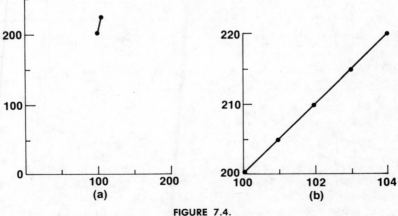

FIGURE 7.4.

Example 7.4 The pressure of a sample of gas maintained at constant volume is measured at a series of temperatures:

P(mm Hg)	90	93	96	99	102
T(°K)	300	310	320	330	340

Plot this data (P on the vertical axis, T on the horizontal axis) on the graph paper below (Figure 7.5).

Solution We must decide what values to assign to the divisions along the two axes. Let us first consider the x axis. We note that there are 20 spaces along the x axis; the temperature range to be covered is 40°K (340°K − 300°K = 40°K). It would seem reasonable to let one space represent 2°K. Starting with 300°K at the far left, successive divisions become:

$$300, 302, 304, 306, 308, 310, \cdots 336, 338, 340$$

Along the y axis, we have 15 spaces which must cover the range between 90 and 102 mm Hg, i.e., 12 mm Hg. To use the

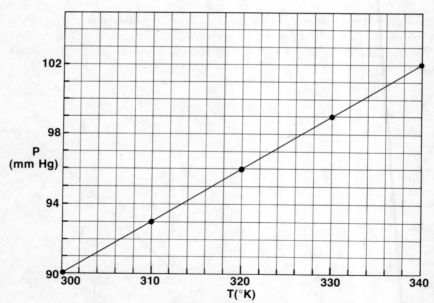

FIGURE 7.5. Plot of pressure of gas sample (mm Hg) vs temperature (°K).

entire distance, we could let one space represent

$$12 \text{ mm Hg}/15 = 0.80 \text{ mm Hg}$$

In practice, this choice would be awkward, because successive divisions would not correspond to integers. That is, we would have:

$$90.0, 90.8, 91.6, 92.4, 93.2, \cdot \cdot \cdot 100.4, 101.2, 102.0$$

It would be simpler to let one space along the y axis represent 1 mm Hg. If we start at the bottom with 90 mm Hg, successive divisions become:

$$90, 91, 92, 93, \cdot \cdot \cdot 99, 100, 101, 102$$

Using 2°K and 1 mm Hg as divisions, the data is plotted in Figure 7.5. Notice that only a relatively small region near the top of the paper is not used.

EXERCISES

1. Graph the following functions, from $x = 0$ to $x = 5$:

   a. $y = 6.0 x$ d. $\log y = 8.00/x - 5.00$

   b. $y = 12/x$ e. $y = -1.6x + 12.0$

   c. $y = 2.10 x^2$ f. $2y = 3x - 9$

2. Which of the functions in number (1) are linear? For each linear function, give the slope and intercept.

3. Graph the following sets of data, drawing a smooth curve through the points:

a.	y	12.8	15.2	17.6	20.0	22.4
	x	0	1	2	3	4
b.	y	6.00	5.00	4.29	3.75	
	x	0.50	0.60	0.70	0.80	
c.	y	10.0	100	450	1300	3200
	x	200	250	300	350	400

7.3   LINEAR FUNCTIONS

As pointed out earlier, several functions that we deal with in general chemistry are of the form:

$$y = ax + b$$

and yield a straight line when y is plotted against x. Many more complex relationships can be transformed into linear functions by a simple change in variables. Consider, for example, the function:

$$y = a/x$$

If we plot y vs x, we get a hyperbola rather than a straight line (Figure 7.2a). However, if we plot y vs $1/x$, we get a straight line with a slope of a (Figure 7.6a). In other words, by changing the independent variable from x to $1/x$, we transform a hyperbolic into a linear function.

Another functional relationship which is readily converted to a linear function is:

$$\log y = \frac{-A}{x} + B$$

If we choose $\log y$ and $1/x$ to be our variables rather than y and x, we obtain a linear function (Figure 7.6b).

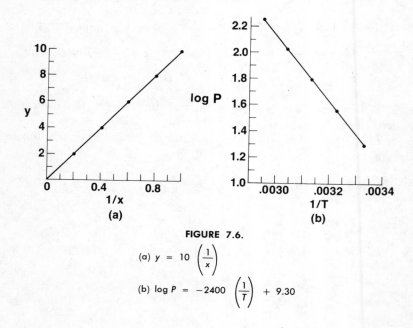

FIGURE 7.6.

(a) $y = 10 \left(\dfrac{1}{x} \right)$

(b) $\log P = -2400 \left(\dfrac{1}{T} \right) + 9.30$

Converting a more complex functional relationship into a linear function has an obvious advantage; it greatly facilitates the processes of interpolation and extrapolation. Suppose, for example, that we wished to use Figure 7.2a to find y when $x = 0.5$. This would be rather difficult to do, since it would require extending the curve in a region where the slope is changing rapidly. On the other hand, it would be relatively easy to extend the straight line in Figure 7.6a to find that:

$$y = 20 \text{ when } x = 0.5 \qquad (1/x = 2.0)$$

Transforming complex functions to linear functions has another, less obvious advantage. It often happens that the slope (and sometimes the intercept) of the resultant straight line has direct physical significance. Consider, for example, the relationship between the vapor pressure of a liquid and the absolute temperature:

$$\log P = \frac{-A}{T} + B \qquad (7.9)$$

If $\log P$ is plotted vs $1/T$, as in Figure 7.6b:

$$\text{slope} = -A \qquad (7.10)$$

Comparing Equations 7.6 and 7.9, we see that:

$$A = \frac{\Delta H_{vap}}{2.3\,R} \qquad (7.11)$$

From Equations 7.10 and 7.11, we deduce that the heat of vaporization of a liquid can be obtained by taking the slope of a plot of $\log P$ vs $1/T$. That is:

$$\Delta H_{vap} = -2.3\,R\,(\text{slope}) \qquad (7.12)$$

Since the slope and the intercept of a linear function are often physically meaningful quantities, it behooves us to consider rather carefully how these quantities can best be determined accurately. The problem is essentially as follows: Suppose we have obtained data in the laboratory which can be fitted, perhaps by transforming variables, to a linear function. How can we best use that data to determine the constants in the equation:

$$y = ax + b$$

where a is the slope and b is the intercept?

The procedure which we follow to obtain the constants a and b depends upon how much data we have available. We can distinguish two different cases.

Case 1. *If only two points are available*, we proceed as follows: Using the subscript $_2$ to refer to the second point and $_1$ to the first point:

$$y_2 = ax_2 + b \tag{7.13}$$

$$y_1 = ax_1 + b \tag{7.14}$$

Subtracting: $y_2 - y_1 = a(x_2 - x_1)$

$$\text{or: } a = \frac{y_2 - y_1}{x_2 - x_1} = \frac{\Delta y}{\Delta x}$$

Having calculated a, b is readily obtained by substituting for a in one of the simultaneous equations, 7.13 or 7.14.

Example 7.5   Consider the equation:

$$\Delta G = \Delta H - T\,\Delta S \qquad (a = -\Delta S, \quad b = \Delta H)$$

If ΔG is found to be 8.4 kcal at 300°K, and 10.8 kcal at 400°K, calculate ΔH and ΔS.

Solution

$$\Delta S = -a = -\frac{(\Delta G_2 - \Delta G_1)}{T_2 - T_1} = \frac{-2.4 \text{ kcal}}{100°\text{K}} = -0.024 \text{ kcal}/°\text{K}$$

To obtain ΔH, we apply the equation: $\Delta G = \Delta H - T\,\Delta S$ at 300°K:

$$8.4 \text{ kcal} = \Delta H - 300°\text{K} (-0.024 \text{ kcal}/°\text{K})$$

$$\Delta H = 8.4 \text{ kcal} - 7.2 \text{ kcal} = 1.2 \text{ kcal}$$

Clearly, calculating the values of a and b from only two data points is, at best, a risky business. If either of the points happens to be seriously in error, the values that we get for the slope and intercept will be highly unreliable.

Case 2. *If three or more points are available*, the constants a and b can be determined by graphing the data, drawing the "best" straight line through the points, and measuring the slope and intercept.

In following this procedure, it is rarely possible to draw a straight line which passes *exactly* through all of the points. At best, some of the points will be above the line and some will be below it. What we attempt to do is to draw the line in such a way that we have about as many points above the line as below. More exactly, we estimate visually the position of the line such that the sum of the distances of points above it will be equal to the sum of the distances of points below the line. A transparent plastic straightedge, triangle, or ruler should be used, so that we can see all the points while we are trying to decide where to draw the line.

Example 7.6 Let us suppose that the data given in Example 7.5 has been extended to give ΔG at four different temperatures:

ΔG	4.4 kcal	6.8 kcal	8.4 kcal	10.8 kcal
T	100°K	200°K	300°K	400°K

Using a graphical method, determine the constants ΔH and ΔS in the equation:

$$\Delta G = \Delta H - T\Delta S$$

Solution In Figure 7.7, we have plotted ΔG vs T; the four data points are shown as small circles. We have attempted to draw the "best" straight line through these points. Notice that it does not pass exactly through the center of any point. The points at 300°K and 100°K fall slightly below the line, while those at 400°K and 200°K are slightly above the line. It appears that the sums of the distances above and below the line are about equal.

The intercept of the line we have drawn is 2.6 kcal. The slope can be found from the values of ΔG and T at the ends of the line:

$$\text{slope} = \frac{10.6 \text{ kcal} - 2.6 \text{ kcal}}{400°K - 0°K} = 0.020 \text{ kcal}/°K$$

We deduce that: $\Delta H = +2.6$ kcal

$$\Delta S = -\text{slope} = -0.020 \text{ kcal}/°K$$

It is interesting to compare these values with those calculated in Example 7.5 ($\Delta H = +1.2$ kcal; $\Delta S = -0.024$ kcal/°K). Clearly, by obtaining two more points, we have considerably revised our estimate of the intercept and slope.

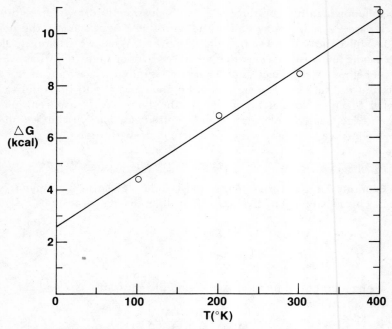

FIGURE 7.7. Relation between ΔG and T: $\Delta G = +2.6 + 0.020T$.

Frequently, experimental data, even though they can in principle be fitted to a linear function, show sufficient scatter to make it difficult to decide how to draw a straight line through the several points. One way to handle this situation is to use a statistical approach known as the method of **least squares.** Without attempting to justify this method mathematically, we shall simply state, without proof, that it gives us the equation of the straight line through the data points which minimizes the standard deviation (Chapter 11) of the observed y values. The method of least squares tells us that, in this sense; the "best" values of the constants a and b are:

$$a = \frac{\Sigma y \, \Sigma x - n\Sigma yx}{\Sigma x \, \Sigma x - n\Sigma x^2} \qquad (7.15)$$

$$b = \frac{\Sigma yx \, \Sigma x - \Sigma y \, \Sigma x^2}{\Sigma x \, \Sigma x - n\Sigma x^2} \qquad (7.16)$$

where: n = no. of points

$\Sigma y = y_1 + y_2 + y_3 + \cdots \cdots = $ sum of all the y values

$\Sigma x = x_1 + x_2 + x_3 + \cdots \cdots = $ sum of all the x values

$\Sigma yx = y_1x_1 + y_2x_2 + y_3x_3 + \cdots = $ sum of the yx products

$\Sigma x^2 = x_1^2 + x_2^2 + x_3^2 + \cdots \cdots = $ sum of the squares of the x values

The application of this method to a quite simple set of data points is illustrated in Example 7.7. It should be pointed out that the tedious arithmetic involved when we have a large number of points can be eliminated by using a computer, which is employed routinely to carry out least-squares calculations (see Chapter 12).

Example 7.7 Apply the method of least squares to the data in Example 7.6 to find the constants in the linear equation:

$$\Delta G = \Delta H - T\Delta S$$

Solution We first compile a table giving values of y, x, yx, and x^2 for each point, realizing that $\Delta G = y$, $T = x$.

y	x	yx	x^2
4.4	100	440	1.00×10^4
6.8	200	1360	4.00×10^4
8.4	300	2520	9.00×10^4
10.8	400	4320	16.00×10^4
30.4	1000	8640	30.00×10^4

Hence: $\quad \Sigma y = 3.04 \times 10^1$; $\quad \Sigma yx = 8.64 \times 10^3$; $\quad n = 4$

$$\Sigma x = 1.00 \times 10^3; \quad \Sigma x^2 = 3.00 \times 10^5$$

Solving for a and b, using Equations 7.15 and 7.16:

$$a = \frac{(3.04 \times 10^1)(1.00 \times 10^3) - 4(8.64 \times 10^3)}{(1.00 \times 10^3)(1.00 \times 10^3) - 4(3.00 \times 10^5)}$$

$$= \frac{-0.416 \times 10^4}{-0.200 \times 10^6} = 2.1 \times 10^{-2}$$

$$b = \frac{(8.64 \times 10^3)(1.00 \times 10^3) - (3.04 \times 10^1)(3.00 \times 10^5)}{(1.00 \times 10^3)(1.00 \times 10^3) - 4(3.00 \times 10^5)}$$

$$= \frac{-0.48 \times 10^6}{-0.20 \times 10^6} = +2.4$$

We deduce that: $\quad \Delta H = +2.4 \text{ kcal}$

$$\Delta S = -a = -0.021 \text{ kcal/}°\text{K}$$

Note that these values compare quite closely with those calculated by a graphical method in Example 7.6 ($\Delta H = +2.6$ kcal, $\Delta S = -0.020$ kcal/$°$K).

The method of least squares is actually a much more powerful technique than this simple example might imply. It can be used not only to find the constants in the linear equation:

$$y = ax + b$$

but also those in the general power series:

$$y = a + bx + cx^2 + dx^3 + \cdots$$

For a discussion of the theory behind the least-squares approach and its general applicability, you may wish to consult one of the references listed in Appendix 1, or either of the following sources: H. W. Salzberg, J. I. Morrow, and S. R. Cohen: *Laboratory Course in Physical Chemistry*. Academic Press, New York, 1966, pp. 23–27.   I. Klotz: *Chemical Thermodynamics*. W. A. Benjamin, Inc., New York, 1964, pp. 25–27.

EXERCISES

1. Transform each of the following into a linear function by an appropriate change of variables:

a. $y = 6x^2 + 1$ d. $\log y = a \log x$
b. $y = 3/x - 4$ e. $y = x^2 - 2x + 1$
c. $\log y = ax^2$ f. $y = x^2 + 3x + 1$

2. Assuming that y is a linear function of x, determine the constants a and b if:

a. $y = 0$ when $x = 0$ and $y = 3$ when $x = 2$
b. $y = 0$ when $x = 3$ and $y = 9$ when $x = 4$
c. $y = 4$ when $x = 2$ and $y = -2$ when $x = 5$

3. Assuming that the data points:

y	5.0	7.8	10.8	14.0	16.8	19.8
x	1.0	2.0	3.0	4.0	5.0	6.0

can be fitted to the equation: $y = ax + b$, find a and b by:

a. A graphical method.
b. The method of least squares.

PROBLEMS

7.1 For the reaction:

$$2\,SO_2(g) + O_2(g) \rightarrow 2\,SO_3(g)$$

$\Delta H = -47.0$ kcal, and $\Delta G = -33.4$ kcal at $300°K$. Using the equation:
$\Delta G = \Delta H - T\Delta S$, calculate:

    a. ΔS

    b. ΔG at $500°K$.

    c. The temperature at which $\Delta G = 0$.

7.2 For the reaction:

$$CO(g) + NO_2(g) \rightarrow CO_2(g) + NO(g)$$

the rate constant, k, is 0.276 at $658°K$ and 14.5 at $783°K$. Using the equation:

$$\log \frac{k_2}{k_1} = \frac{\Delta E_{act}}{2.30\ R} \left[\frac{T_2 - T_1}{T_2 T_1} \right]$$

calculate:

    a. The energy of activation, ΔE_{act} ($R = 1.99$ cal/$°K$).

    b. The rate constant at $700°K$.

7.3 Using the information given in Problem 7.1, plot ΔG vs T from $T = 0$ to $T = 500°K$.

7.4 The following data were obtained for the concentrations of Ag^+ and CrO_4^{2-} in a solution saturated with Ag_2CrO_4:

conc. CrO_4^{2-}	1.0×10^{-4}	0.25×10^{-4}	0.11×10^{-4}	0.067×10^{-4}
conc. Ag^+	1.0×10^{-4}	2.0×10^{-4}	3.0×10^{-4}	4.0×10^{-4}

    a. Plot the concentration of CrO_4^{2-} (y) vs the concentration of Ag^+ (x).

    b. Plot the concentration of CrO_4^{2-} vs $1/$(the concentration of $Ag^+)^2$.

    c. Write the equation relating the concentration of CrO_4^{2-} to the concentration of Ag^+.

7.5 Consider the ideal gas law: $PV = nRT$
For one mole of an ideal gas, plot:

    a. P vs T at $V = 10$ liters   ($R = 0.0821$ lit atm/mole$°K$).

    b. PV vs P at $T = 298°K$.

    c. PV vs T.

7.6 For each of the following functional relationships, indicate what quantities should be plotted against each other to obtain a straight line:

a. $P = \dfrac{\text{constant}}{V}$

d. $\log K_p = \dfrac{-\Delta H}{2.30\, RT} + \text{constant}$

b. $m = \dfrac{\text{constant}}{u^2}$

e. $E = E° - 0.059 \log (\text{conc. } H^+)^2$

c. $\Delta G° = -1373 \log K_p$

f. $\log (\text{conc } A) = kt + \text{constant}$

7.7 For the radioactive decomposition of radon, the following data are obtained for the amount of radon (A) remaining as a function of time (t), expressed in days:

A	10.0	8.2	7.0	5.8	5.0	4.2
t	0	1	2	3	4	5

It is believed that the data fit the equation:

$$\log A = at + b$$

Calculate the constants a and b:
  a. Using the two points for $t = 0$ and $t = 4$.
  b. By plotting $\log A$ vs t and drawing the best straight line through the points.
  c. By applying the method of least squares to the data.

7.8 For a water solution of acetic acid, the following equation applies:

$$pH = a \log (\text{conc. HA}) - \frac{1}{2} \log K_a$$

Using the following data:

pH	2.4	2.9	3.4	3.8
conc. HA	1.00	0.100	0.0100	0.00100

Calculate K_a by:
  a. Plotting pH vs log (concentration of HA) and determining the intercept.
  b. Applying the method of least squares.

7.9 The following data were obtained for the dependence of the vapor pressure of chloroform on temperature:

P	61.0	100.5	159.6	246.0
T	273	283	293	303

Use a graphical method to calculate the heat of vaporization, using the equation:

$$\log P = \frac{-\Delta H_{vap}}{(2.30)(1.99)\,T} + \text{constant}$$

7.10 For a certain reaction, the following data were obtained for the variation of the concentration of reactant with time:

conc.	1.00	0.80	0.63	0.50	0.40	0.32
t (min)	0	1	2	3	4	5

This reaction may be:

"Zero order," in which case $C_0 - C = kt$

"First order," in which case $\log \dfrac{C_0}{C} = \dfrac{kt}{2.3}$

"Second order," in which case $\dfrac{1}{C} - \dfrac{1}{C_0} = kt$

where C_0 is the original concentration, C is the concentration at time t, and k is the rate constant. Using a graphical method, determine the order of the reaction and calculate k.

SPACE GEOMETRY AND TRIGONOMETRY

In general chemistry, we are concerned with the geometry of molecules, complex ions, and crystals. To understand the properties of these polyatomic species, we need to be familiar with certain principles of trigonometry and solid geometry. In this chapter, we shall deal with selected topics in these fields of mathematics, restricting our discussion to concepts which are particularly relevant to general chemistry.

8.1 TRIGONOMETRIC FUNCTIONS

The two fundamental trigonometric functions are the sine (sin) and cosine (cos). These functions can be defined with reference to a right triangle (Figure 8.1):

$$\sin A = \frac{\text{opposite side}}{\text{hypotenuse}} = \frac{a}{c} \tag{8.1}$$

$$\cos A = \frac{\text{adjacent side}}{\text{hypotenuse}} = \frac{b}{c} \tag{8.2}$$

Two other trigonometric functions, which we shall have less occasion to use, are the tangent (tan) and cotangent (cot):

$$\tan A = \frac{\text{opposite side}}{\text{adjacent side}} = \frac{a}{b} ; \quad \cot A = \frac{\text{adjacent side}}{\text{opposite side}} = \frac{b}{a}$$

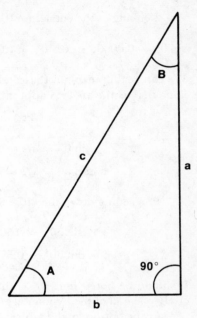

FIGURE 8.1. $\sin A = \dfrac{a}{c}$; $\cos A = \dfrac{b}{c}$.

Certain relationships between these functions should be obvious from their definitions. Thus, we have:

$$\tan A = \frac{a}{b} = \frac{a/c}{b/c} = \frac{\sin A}{\cos A}$$

$$\cot A = \frac{b}{a} = \frac{1}{a/b} = \frac{1}{\tan A}$$

The relationship between the sine of an angle and its cosine:

$$\sin^2 A + \cos^2 A = 1 \tag{8.3}$$

is less obvious. It can be derived (see Exercise number 2 at the end of this section) from the Pythagorean theorem, which relates the lengths of the sides of a right triangle:

$$a^2 + b^2 = c^2 \tag{8.4}$$

The table of trigonometric functions given in Appendix 2 can be used to find the sine, cosine, tangent, or cotangent of any angle between 0° and 180°. We shall now consider how to use this table in the three regions, 0 to 45°, 45 to 90°, and 90 to 180°.

0 to 45°.   *Read the value of the function (listed at the top of each column) opposite the angle listed in the column at the far left.*

Looking at the entries for 0°, we have:

$$\sin 0° = 0.0000; \quad \cos 0° = 1.0000; \quad \tan 0° = 0.0000; \quad \cot 0° = \infty$$

We can readily explain these values by referring to Figure 8.1, and realizing that as the angle A approaches 0°:

$$a \rightarrow 0, \quad \text{and} \quad b \rightarrow c$$
$$\sin 0° = a/c = 0; \quad \cos 0° = b/c = 1$$
$$\tan 0° = a/b = 0; \quad \cot 0° = b/a = \infty$$

Proceeding down the table, we find that at 30°:

$$\sin 30° = 0.5000; \quad \cos 30° = 0.8660;$$
$$\tan 30° = 0.5774; \quad \cot 30° = 1.732$$

It should be noted that 30° is the angle at which the opposite side is precisely one half of the hypotenuse (Figure 8.2a).

At the bottom of the table, we have:

$$\sin 45° = \cos 45° = 0.7071; \quad \tan 45° = \cot 45° = 1.0000$$

At 45°, the opposite and adjacent sides are exactly equal in length (Figure 8.2b).

45 to 90°. To find trigonometric functions in this range, we take advantage of the fact that:

$$\sin (90° - A) = \cos A$$
$$\cos (90° - A) = \sin A$$
$$\tan (90° - A) = \cot A$$
$$\cot (90° - A) = \tan A$$

(These relationships are readily proved using Figure 8.1 and realizing that $B = 90° - A$.)

For convenience, the table of trigonometric functions is supplied with a right hand index which allows us to perform these operations automatically. *For angles between 45° and 90°, we use the right hand index in conjunction with the function listed at the bottom of each column.* At 60°, we find:

$$\sin 60° = 0.8660; \quad \cos 60° = 0.5000;$$
$$\tan 60° = 1.732; \quad \cot 60° = 0.5774$$

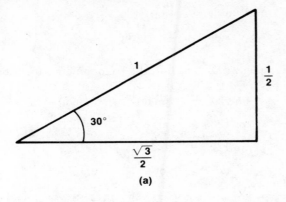

(a)

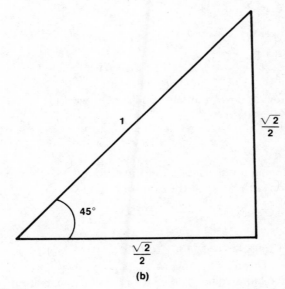

(b)

FIGURE 8.2.

(a) $\sin 30° = \dfrac{1/2}{1} = 0.5000;$ $\cos 30° = \dfrac{\sqrt{3}/2}{1} = \dfrac{1.7320}{2} = 0.8660$

(b) $\sin 45° = \cos 45° = \dfrac{\sqrt{2}/2}{1} = \dfrac{1.4142}{2} = 0.7071$

Compare these values with the corresponding functions for 30° listed previously.

90 to 180°.   Here we use the relationships:

$$\sin A = \sin(180° - A)$$
$$\cos A = -\cos(180° - A)$$

$$\tan A = -\tan(180° - A)$$

$$\cot A = -\cot(180° - A)$$

Suppose, for example, we wish to find the trigonometric functions of 120°:

$$\sin 120° = \sin 60° = 0.8660$$

$$\cos 120° = -\cos 60° = -0.5000$$

$$\tan 120° = -\tan 60° = -1.732$$

$$\cot 120° = -\cot 60° = -0.5774$$

Example 8.1　Find:
  a. sin 18°
  b. cos 59°
  c. tan 109°

Solution
  a. Moving down the left hand index, we find in the column headed "sin":

$$\sin 18° = 0.3090$$

  b. Moving up the right hand index to 59°, we find in the column with "cos" at the bottom:

$$\cos 59° = 0.5150$$

  c. $\tan 109° = -\tan(180° - 109°) = -\tan 71°$
      To find the tangent of 71°, we move up to 71° on the right hand index and read across to the column labeled "tan" at the bottom:

$$\tan 71° = 2.904; \qquad \tan 109° = -2.904$$

The slide rule (Chapter 3) can be used to obtain sines and tangents of angles, using the S and T scales respectively. (Many slide rules also have an "ST" scale, which can be used to find either sines or tangents below 0.1000, corresponding to angles less than 5.7°.) Having found the sine and tangent of an angle, it is always possible to calculate its cosine:

$$\cos A = \sin A/\tan A$$

Trigonometric functions are of particular use to us in the area of molecular structure. Frequently, we know certain distances and/or angles between bonded atoms and are required to find other distances in the molecule. In many cases, we can do this by using the trigonometric functions directly, in conjunction with the properties of right triangles (Examples 8.2 and 8.3).

Example 8.2 In the H_2O molecule, the $\angle HOH$ is $104°$; each H atom is located 0.958 Å from the central oxygen atom (Figure 8.3). What is the distance between the two hydrogen atoms?

Solution A simple way to analyze this problem is indicated in Figure 8.3. The dotted line drawn perpendicular to the H—H axis bisects the $\angle HOH$. It follows that:

$$A = 52°; \qquad \sin A = 0.788$$

But: $\sin A = a/c; \qquad a = c \sin A$
$$= (0.958 \text{ Å})(0.788) = 0.755 \text{ Å}$$

From the figure, it is clear that the distance between the two hydrogen atoms is $2a$, or 1.51 Å.

Example 8.3 The compound N_2F_2 exists as a mixture of two forms, referred to as *cis* and *trans* (Figure 8.4). In both isomers, the N—N distance is 1.25 Å, the N—F distance is 1.44 Å, and the $\angle NNF$ is $115°$. Calculate the distance between fluorine atoms in the *cis* isomer.

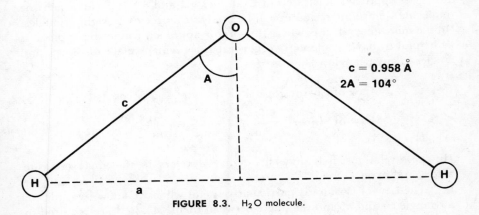

FIGURE 8.3. H_2O molecule.

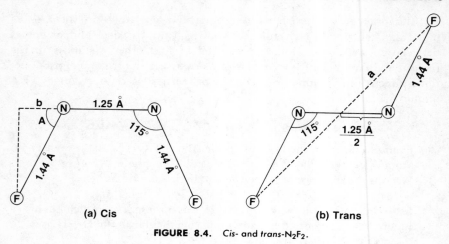

FIGURE 8.4. *Cis- and trans-*N_2F_2.

Solution   Referring to Figure 8.4a, we note that the angle A must be $180° - 115° = 65°$. It is possible to solve for the distance b by using the defining equation for the cosine:

$$\cos A = \frac{b}{1.44 \text{ Å}}: \qquad b = (1.44 \text{ Å})(\cos 65°)$$

$$= (1.44 \text{ Å})(0.423) = 0.609 \text{ Å}$$

It should be clear from the figure that:

$$\text{F—F distance} = \text{N—N distance} + 2b$$

$$= 1.25 \text{ Å} + 1.22 \text{ Å} = 2.47 \text{ Å}$$

The approach illustrated in Examples 8.2 and 8.3 is useful when the molecule we are interested in can be analyzed in terms of right triangles. In the more general case, where this simple approach cannot be applied, it is helpful to have available two general rules which relate distances and angles in any triangle. These are:

Law of Cosines: $a^2 = b^2 + c^2 - 2\,bc \cos A$ \qquad (8.5)

Law of Sines: $\dfrac{a}{\sin A} = \dfrac{b}{\sin B} = \dfrac{c}{\sin C}$ \qquad (8.6)

(In these relations, it is understood that A refers to the angle opposite side a, and so on.)

The Law of Cosines is particularly useful when we know two sides of a triangle (b and c) and the angle between them (A), and are required to find the length of the third side. This situation arises quite frequently in problems dealing with molecular structure (Examples 8.4 and 8.5).

Example 8.4 In the nitrosyl chloride molecule (Figure 8.5), the $\angle ONCl$ is 113°, the N—O bond distance is 1.14 Å, and the N—Cl bond distance is 1.97 Å. Calculate the O—Cl distance.

Solution Taking a to be the O—Cl distance, we have:

$$a^2 = (1.14 \text{ Å})^2 + (1.97 \text{ Å})^2 - 2(1.14 \text{ Å})(1.97 \text{ Å}) \cos 113°$$

$$= 1.30 (\text{Å})^2 + 3.88 (\text{Å})^2 - 4.49 (\text{Å})^2 (-0.391)$$

$$= 6.94 \text{ (Å)}^2$$

$$a = 2.63 \text{ Å}$$

Example 8.5 Referring back to Example 8.3 and Figure 8.4, calculate the distance between the fluorine atoms in the *trans* isomer of N_2F_2.

Solution This problem is readily solved if we realize that the line joining the two fluorine atoms must pass midway between the two nitrogen atoms (Figure 8.4b). Applying the Law of Cosines, we have:

$$a^2 = \left(\frac{1.25 \text{ Å}}{2}\right)^2 + (1.44 \text{ Å})^2 - 2\left(\frac{1.25 \text{ Å}}{2}\right)(1.44 \text{ Å}) \cos 115°$$

$$= 0.39 (\text{Å})^2 + 2.07 (\text{Å})^2 + (1.25 \text{ Å})(1.44 \text{ Å})(0.423)$$

$$= 3.22 (\text{Å})$$

$$a = 1.80 \text{ Å}$$

$$\text{F—F distance} = 2(1.80 \text{ Å}) = 3.60 \text{ Å}$$

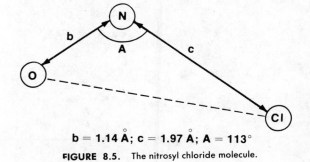

b = 1.14 Å; c = 1.97 Å; A = 113°

FIGURE 8.5. The nitrosyl chloride molecule.

EXERCISES

1. Find the values of the following trigonometric functions:
   a. sin 29°   d. tan 49°
   b. cos 18°   e. sin 112°
   c. sin 65°   f. cos 95°

2. Use the Pythagorean theorem (8.4) in conjunction with Equations 8.1 and 8.2 to prove:

$$\sin^2 A + \cos^2 A = 1$$

3. The H_2S molecule resembles the H_2O molecule shown in Figure 8.3, except that $c = 1.33$ Å and $2A = 92°$. Find the H—H distance.

4. The SCl_2 molecule is bent with $\angle ClSCl = 103°$. The S—Cl bond distance is 2.00Å. Calculate the Cl—Cl distance, using the Law of Cosines.

5. Three atoms with radii of 1.40Å, 1.70Å, and 2.00Å, respectively, are arranged in a triangular pattern so that each atom touches two others. Calculate the angles between the lines joining the centers of the three atoms.

8.2 MENSURATION FORMULAE OF PLANE FIGURES

In general chemistry, we may have occasion to use the relationships listed in Table 8.1 to calculate areas of various two-dimensional figures.

Table 8.1 Area Formulae

Figure	Area
1. Triangle	$= bh/2$ (b = base, h = length of perpendicular to b from opposite vertex)
	$= \dfrac{ab \sin C}{2}$ (a, b, and c are sides; C is the angle opposite side c)
2. Rectangle	$= ab$ (a and b are sides)
3. Square	$= a^2$ (a is the length of a side)
4. Any regular polygon	$= \dfrac{na^2}{4} \cot \dfrac{180}{n}$ (n = no. of sides, a = length of side)
5. Circle	$= \pi r^2$ (r = radius)

Another useful relationship gives the sum of the interior angles of a polygon of n sides:

$$(n - 2)\ 180° \tag{8.7}$$

For a regular polygon (all sides equal), each interior angle is:

$$\frac{(n - 2)}{n}\ 180° \tag{8.8}$$

Example 8.6 In the benzene molecule, C_6H_6, the carbon atoms are located at the vertices of a regular hexagon; the C—C bond distance is 1.40 Å. Find:
a. The $\angle CCC$.
b. The area enclosed by the carbon atoms.

Solution
a. Using Equation 8.8, we figure out that each interior angle in a regular polygon of 6 sides is:

$$\frac{4}{6} \times 180° = 120°$$

b. Using relation (4) of Table 8.1: area $= \dfrac{na^2}{4} \cot \dfrac{180°}{n}$

$$= \frac{6}{4}\ (1.40\ Å)^2 \cot 30°$$

$$= \frac{6}{4}\ (1.40\ Å)^2 (1.732)$$

$$= 5.09\ (Å)^2$$

EXERCISES

1. The normal angle between carbon atoms, i.e., $\angle CCC$, is 109°. Consider cyclic compounds in which the carbon atoms are located at the corners of an equilateral triangle, a square, a regular pentagon, and a regular hexagon. Which type of compound would you expect to be most stable, from the point of view of bond angles?

2. A small crystal has the shape of a prism. Each of the four faces of the prism is an equilateral triangle whose side has a length of 1.20 mm. Calculate the total area, in cc, of the faces of the crystal.

8.3 THREE-DIMENSIONAL FIGURES

The Sphere. When we refer to the radii of individual atoms or monatomic ions, it is understood that we are considering the atom or ion to be a sphere. For this reason, among others, we may find useful the relationships listed in Table 8.2.

Table 8.2 Properties of a Sphere of Radius r

Volume $= \dfrac{4}{3} \pi r^3$

Surface area $= 4\pi r^2$

Volume of segment of height $h = \dfrac{\pi h^2}{3} (3r - h)$

Area of segment of height $h = 2\pi rh$

Example 8.7 Calculate the volume, in cc, of 6.02×10^{23} copper atoms, given that the atomic radius of copper is 1.28 Å.

Solution Since 1 Å $= 10^{-8}$ cm, $r = 1.28 \times 10^{-8}$ cm

Volume of copper atom $= \dfrac{4}{3} \pi (1.28 \times 10^{-8} \text{ cm})^3$

$$= 8.79 \times 10^{-24} \text{ cc}$$

Volume of 6.02×10^{23} copper atoms $=$
$$(6.02 \times 10^{23})(8.79 \times 10^{-24} \text{ cc})$$

$$= 5.29 \text{ cc}$$

The Cube. The cube is a basic structural unit of a great many crystals. A knowledge of the geometry of the cube is of particular importance in calculations dealing with the so-called "unit cell" of a crystal. (The unit cell is the simplest structural unit which, repeated over and over again in three dimensions, will generate the crystal.) Perhaps the simplest unit cell would be a cube in which the atoms at the corner of the cube touch each other. A much more common unit cell is the *body-centered cube*, which refers to a cube with atoms at each corner and one in the center. In a body-

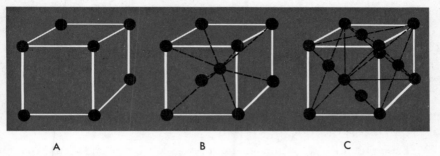

FIGURE 8.6.   Cubic packing. A, simple; B, body-centered; C, face-centered.

centered cubic unit cell, the atom at the center touches those at the corners. Another common unit cell is the *face-centered cube*, in which there is an atom at each corner and one in the center of each face. In such a cell, atoms touch along a face diagonal.

Example 8.8   In a simple cubic unit cell, in which atoms at each corner touch each other, what fraction of the total volume is empty space?

Solution   We have a total of 8 atoms at the corners of the unit cell. However, we must take into account the fact that these cubes are stacked in three dimensions. A total of eight different cubes touch at a corner and possess an equal share of the atom located at that point. In effect, only $\frac{1}{8}$ of that atom can be assigned to a particular cube. This means that we have:

$$8 \times \frac{1}{8} = 1 \text{ atom per cube}$$

If we let r represent the radius of the atom, its volume is:

$$V \text{atom} = \frac{4}{3} \pi r^3$$

To find the volume of the cube itself, we note that since two atoms touch along one edge of the cube, the length of one side must be:

$$a = r + r = 2r$$

The volume of the cube is:

$$V \text{cube} = (2r)^3 = 8 r^3$$

The fraction of the total volume which is filled is:

$$\frac{V \text{ atom}}{V \text{ cube}} = \frac{\frac{4}{3}\pi r^3}{8r^3} = \frac{\pi}{6} = 0.524$$

Fraction of empty space $= 1 - 0.524 = 0.476$

Example 8.9 A certain metal crystallizes in a face-centered cubic structure. The side of the cube which forms the unit cell has a length of 2.96 Å. Calculate the atomic radius of the metal.

Solution We start by realizing that atoms are touching along the diagonal which passes across the face of the cube. The length of that diagonal, according to Table 8.3, is:

$$a\sqrt{2} = (2.96 \text{ Å})(1.41) = 4.17 \text{ Å}$$

But, the length of the diagonal must be equivalent to four atomic radii (Figure 8.7). Hence:

$$4r = 4.17 \text{ Å}$$
$$r = 1.04 \text{ Å}$$

Table 8.3 Properties of a Cube of Side a

Volume $= a^3$
Surface area $= 6\,a^2$
Length of face diagonal $= a\sqrt{2}$
Length of diagonal through center $= a\sqrt{3}$

The Tetrahedron. In many molecules and complex ions, a central atom is bonded to four other atoms. When this happens, it is usually found that the four bonds are directed toward the corners of a *regular tetrahedron*. An example of a tetrahedral species is the methane molecule shown in Figure 8.8. The heavy lines represent the bonds joining carbon to hydrogen; the sides of the regular tetrahedron are outlined by the light lines. Other species showing this geometry include the CCl_4 molecule, the SO_4^{2-} ion, and the $Zn(NH_3)_4^{2+}$ ion.

A regular tetrahedron is a figure which has four faces, all of which are

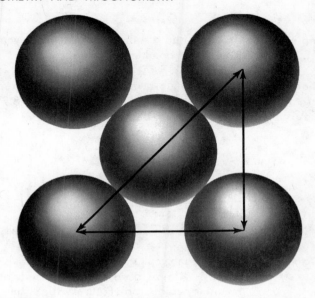

FIGURE 8.7. Atoms touching along diagonal in face-centered cube.

equilateral triangles. Many of the properties of species made up of tetra-hedrally directed bonds can be deduced from the simple trigonometry associated with triangles. In this connection, it is important to know the magnitude of the "tetrahedral angle." This is the angle between two lines drawn from vertices of a tetrahedron to its center. In terms of molecular geometry, it is the angle between two bonds directed toward the corners of a regular tetrahedron (∠ HCH in Figure 8.8):

$$\text{tetrahedral angle } = 109.5° \qquad (8.9)$$

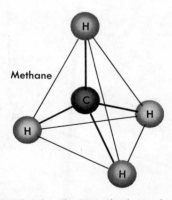

FIGURE 8.8. The CH_4 molecule (tetrahedral).

Example 8.10 Calculate the distance between hydrogen atoms in CH_4, given that the C—H bond length is 1.09 Å.

Solution Consider the triangle formed by the carbon atom in the center and two hydrogen atoms at the vertices (Figure 8.9). The angle A is one half of 109°, or approximately 55°.

$$\sin A = \frac{a}{1.09 \text{ Å}}; \qquad a = (1.09 \text{ Å})(\sin 55°)$$

$$= (1.09 \text{ Å})(0.819) = 0.893 \text{ Å}$$

$$\text{H—H distance} = 2(0.893 \text{ Å}) = 1.79 \text{ Å}$$

The Octahedron. In many molecules and complex ions, a central atom is bonded to six other atoms. Almost invariably, the bonds are directed toward the corners of a regular octahedron (Figure 8.10a). The SF_6 molecule and the complex ions formed by Cr^{3+}, Co^{3+}, and many other transition metal ions show this geometry.

A regular octahedron is a figure with eight faces, all of which are equilateral triangles. The six vertices of a regular octahedron, at which the six fluorine atoms in sulfur hexafluoride are located, are at an equal

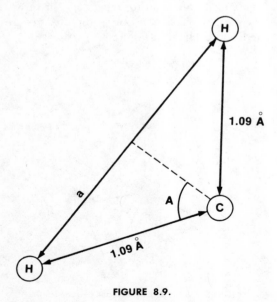

FIGURE 8.9.

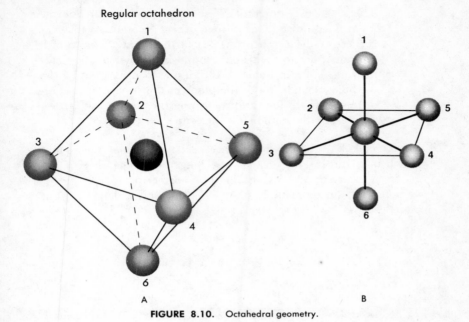

FIGURE 8.10. Octahedral geometry.

distance from the center. One can consider an octahedron as two square pyramids fused together at their bases. (Each pyramid has four faces which are equilateral triangles; the base of the pyramid is outlined by atoms 2, 3, 4 and 5 in Figure 8.10a.) We often take advantage of this feature in representing the structure of species which show octahedral geometry (Figure 8.10b).

Certain characteristics of species with octahedral geometry are illustrated by Example 8.11.

Example 8.11 In the SF_6 molecule, which has the octahedral structure indicated in Figure 8.10, what is:

    a. The $\angle FSF$?

    b. The F—F distance? (S—F distance = 1.58 Å)

Solution

    a. It should be clear from Figure 8.10 that *all* of the F—S—F bond angles are 90°. Remember that the plane figure outlined in Figure 8.10b is a square; the diagonals of a square intersect at an angle of 90°. The two fluorine atoms located above and below the square are on a line perpendicular to the square through its center.

b. There are two different F—F distances. Each fluorine
atom is equidistant from four others located at adjacent
corners of the octahedron, but farther away from a fifth
fluorine atom located at the opposite corner of the octahedron. Thus, in Figure 8.10b, if one focuses attention on
the fluorine atom above the plane of the square, it is equidistant from the four atoms forming the square but farther
away from the atom below the center of the square.

To calculate these two distances, let us consider the square
drawn in Figure 8.11. Clearly, the longer distance, F^1—F^4, is
$2(1.58$ Å$) = 3.16$ Å. To obtain the shorter distance, F^1—F^2,
we note that, since the three points F^1, F^2, and S define a right
triangle:

$$(F^1—F^2 \text{ dist.})^2 = (1.58 \text{ Å})^2 + (1.58 \text{ Å})^2$$
$$= 4.99 \text{ (Å)}^2$$

$$F^1—F^2 \text{ dist.} = 2.23 \text{ Å}$$

The feature of octahedral geometry illustrated in Example 8.11 explains the existence of geometrical isomers with certain complex ions. The
ion $Co(NH_3)_4Cl_2{}^+$ can exist in either the *cis* form, where the two chloride
ions are at adjacent corners of the octahedron, or as the *trans* form, where
the two chloride ions are as far apart as possible, at opposite corners of the
octahedron.

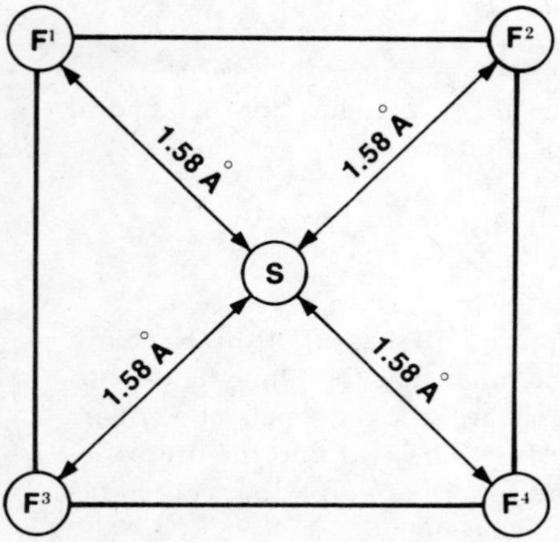

FIGURE 8.11.

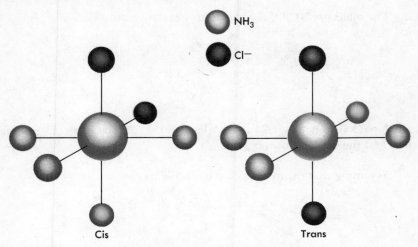

FIGURE 8.12. Isomers of $[Co(NH_3)_4Cl_2]^+$.

EXERCISES

1. Suppose two spheres of equal radius, r, overlap so that the distance between their centers is $1.6r$. Using Table 8.2, calculate the fraction of the total volume of the spheres that lies in the region of overlap.

2. Calculate the fraction of empty space in a unit cell with a body-centered cubic structure and compare it to that calculated for a simple cubic structure (Example 8.8).

3. A certain metal crystallizes in a face-centered cubic structure. The edge of the unit cell is 2.90 Å. Calculate the atomic radius of the metal.

4. Consider a molecule, Ab_4c_2, which has an octahedral structure with atom A in the center and b and c atoms at each vertex. Each A—b distance is 2.00 Å, while each A—c distance is 1.50 Å. Determine all the b—b, c—c, and b—c distances in the *trans* isomer (c atoms far apart).

PROBLEMS

8.1 In the molecule OClO, the $\angle OClO$ is 116°; both Cl—O distances are 1.49 Å. Calculate the distance between the Cl atoms.

8.2 In the ONBr molecule, the $\angle ONBr$ is 117°; the O—N distance is 1.15 Å and the N—Br distance is 2.14 Å. Calculate the Br—O distance.

8.3   The molecule $NClO_2$ has the following structure:

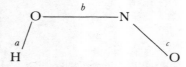

Each N—O distance is 1.24 Å.   Calculate the $\angle ClNO$, the O—O distance, and the O—Cl distance.

8.4   Assuming that the nitrous acid molecule has the structure:

With a = 0.98 Å, b = 1.46 Å, c = 1.20 Å, $\angle ONO$ = 116°, $\angle NOH$ = 104°, calculate the distance from the hydrogen atom to the oxygen atom at the other end of the molecule.

8.5   A square piece of metal foil with a surface area of one cm^2 is covered with a monolayer of argon, which has an atomic radius of 1.54 Å. If one assumes that the argon atoms are lined up as:

how many argon atoms are there at the surface?   What fraction of the surface is actually covered by argon atoms?

8.6   In the arrangement shown in Problem 8.5, what would be the radius of an atom which would just fit into the space between touching argon atoms?

8.7   One mole of zinc atoms (6.02×10^{23} atoms) weighs 65.4 g. If the atomic radius of zinc is 1.33 Å, what is the density of a zinc atom?   The density of zinc metal is about 7.1 g/cc. What fraction of the volume of a zinc crystal is actually occupied by zinc atoms?

8.8   Calculate the fraction of empty space in a face-centered cubic unit cell.

8.9 A certain metal crystallizes with a cubic unit cell 3.00 Å on an edge. What is the atomic radius of the metal if the cell is:

      a. A simple cube?
      b. A body-centered cube?
      c. A face-centered cube?

8.10 In the methane molecule, what is the distance from the carbon atom to the center of the equilateral triangle formed by three hydrogen atoms? (C—H bond distance = 1.09 Å.)

8.11 In the $PtCl_6^{2-}$ ion, the platinum atom is at the center of a regular octahedron with Cl^- ions at each vertex. The Pt—Cl distance is 2.33 Å. What are the Cl—Cl distances?

8.12 How many geometric isomers are there in octahedral species which have the formulae:

      a. Ab_5c d. $Abcd_4$
      b. Ab_3c_3 e. $Abcdefg$
      c. $Ab_2c_2d_2$

      (In each case, the A atom is at the center of the octahedron.)

DIFFERENTIAL CALCULUS

In Chapter 7, we discussed the functional dependence of one variable, y, upon another variable, x. Many times, we are interested in the rate of change of y with respect to x. We may, for example, wish to know how the concentration of a species taking part in a reaction changes with time. Problems of this type are best treated by a branch of mathematics known as differential calculus, which is concerned with the effect that a small change in one variable, x, has upon the value of another variable, y. We shall restrict our discussion of differential calculus to those aspects of the discipline which are most relevant to general chemistry.

9.1 THE MEANING OF THE DERIVATIVE

To illustrate how the concept of a derivative arises, let us consider a problem typical of those encountered in chemical kinetics. In studying the rate of the reaction:

$$N_2O_4(g) \rightarrow 2\,NO_2(g)$$

we might accumulate the following data for the concentration of NO_2 as a function of time:

conc. NO_2	0.00	0.80	1.28	1.57	1.74	1.81
time (hrs)	0	1	2	3	4	5

How can we use this data to determine how rapidly the concentration of NO_2 is increasing at the instant the reaction starts, that is, at zero time? One way to do this would be to plot the data and attempt to draw a tan-

122

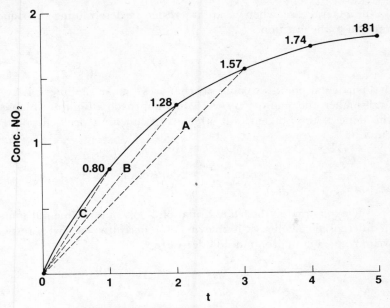

FIGURE 9.1. Rate of formation of NO_2.

gent to the curve at $t = 0$. The slope of this tangent is the quantity we are looking for; *the instantaneous rate of change of concentration* at zero time.

As we can see from Figure 9.1, it is not easy to find the slope of the tangent at $t = 0$. The curvature of the plot makes it difficult to decide where to draw a line which barely touches the curve at this point. On the other hand, it is relatively easy to determine the slopes of successive *secants* drawn from $t = 0$ so as to intersect the curve at $t = 3$, $t = 2$, and $t = 1$, respectively.

$$\text{Secant A:} \quad \text{slope} = \frac{\Delta \text{conc. } NO_2}{\Delta t} = \frac{1.57}{3} = 0.52$$

$$\text{Secant B:} \quad \text{slope} = \frac{\Delta \text{conc. } NO_2}{\Delta t} = \frac{1.28}{2} = 0.64$$

$$\text{Secant C:} \quad \text{slope} = \frac{\Delta \text{conc. } NO_2}{\Delta t} = \frac{0.80}{1} = 0.80$$

It seems reasonable to believe that if we made the interval Δt smaller and smaller, the slope of the secant would approach more and more closely the slope of the tangent. In this particular case, extrapolating from the slopes of the secants A, B, and C, we would estimate:

$$\text{slope of tangent} \approx 1.0$$

In the general case, when we are interested in determining the slope of the tangent to the function:

$$y = f(x)$$

at a particular point, we could say that as we make the interval Δx smaller and smaller, the quantity $\Delta y / \Delta x$ would approach a limiting value equal to the slope of the tangent. In other words, at any specified point on the curve:

$$\underset{\Delta x \to 0}{\text{Limit}} \frac{\Delta y}{\Delta x} = \text{slope of tangent} \tag{9.1}$$

The quantity on the left of Equation 9.1 is of fundamental importance in differential calculus. It is known as the derivative of y with respect to x, written dy/dx, and pronounced "dee y dee x".

$$\frac{dy}{dx} = \underset{\Delta x \to 0}{\text{Limit}} \frac{\Delta y}{\Delta x} \tag{9.2}$$

We see then that the derivative of y with respect to x is simply the slope of the tangent to the curve of y vs x. (This is sometimes referred to simply as "the slope of the curve.")

9.2 DERIVATIVES OF SIMPLE FUNCTIONS

The basic objective of differential calculus can be stated as follows: Given the functional relationship between y and x, find the derivative, dy/dx, at any desired value of x. To show how this can be accomplished, consider the simple function:

$$y = ax \tag{9.3}$$

If we change the value of the independent variable from x to $(x + \Delta x)$, the dependent variable will change from y to $(y + \Delta y)$. But, from Equation 9.3:

$$y + \Delta y = a(x + \Delta x) \tag{9.4}$$

Subtracting Equation 9.3 from Equation 9.4:

$$\Delta y = a \Delta x; \qquad \frac{\Delta y}{\Delta x} = a \tag{9.5}$$

Equation 9.5 tells us that the quotient $\Delta y / \Delta x$ is always equal to the constant, a, regardless of how small Δx may be. Therefore:

$$\underset{\Delta x \to 0}{\text{Limit}} \frac{\Delta y}{\Delta x} = \frac{dy}{dx} = a \tag{9.6}$$

We deduce that, for this function, dy/dx is a constant, a. This should come as no surprise; we recall from Chapter 7 that the equation $y = ax$ gives a straight line with a constant slope of a.

Example 9.1 Show that for the general linear equation:

$$y = ax + b$$

$dy/dx = a$.

Solution Since

$$y = ax + b \tag{1}$$

$$y + \Delta y = a(x + \Delta x) + b = ax + a\Delta x + b \tag{2}$$

Subtracting (1) from (2):

$$\Delta y = a\,\Delta x; \qquad \Delta y / \Delta x = a$$

$$\frac{dy}{dx} = \underset{\Delta x \to 0}{\text{Limit}} \frac{\Delta y}{\Delta x} = a$$

As a slightly more complicated example, consider the problem of evaluating dy/dx for the function:

$$y = ax^2 \tag{9.7}$$

Replacing x by $(x + \Delta x)$, we obtain the value of $(y + \Delta y)$:

$$y + \Delta y = a(x + \Delta x)^2 = ax^2 + 2ax\Delta x + a(\Delta x)^2 \tag{9.8}$$

Subtracting Equation 9.7 from Equation 9.8:

$$\Delta y = 2ax\Delta x + a(\Delta x)^2$$

or:

$$\frac{\Delta y}{\Delta x} = 2ax + a\Delta x \tag{9.9}$$

But, as Δx approaches 0, so does the term $(a\Delta x)$; it becomes vanishingly small compared to the term $2ax$. So:

$$\underset{\Delta x \to 0}{\text{Limit}} \frac{\Delta y}{\Delta x} = \frac{dy}{dx} = 2ax \qquad (9.10)$$

By a process entirely analogous to that just described, it is possible to show that:

$$\text{if } y = ax^3, \text{ then } \frac{dy}{dx} = 3ax^2$$

or, in general:

$$\text{if } y = ax^n, \text{ then } \frac{dy}{dx} = nax^{(n-1)} \qquad (9.11)$$

where n is any number.

Example 9.2   Evaluate dy/dx at $x = 2$ for:
  a. $y = 3x^5$
  b. $y = 3$
  c. $y = 2/x$

Solution   Using Equation 9.11, we have:

  a. $\dfrac{dy}{dx} = (5)(3)x^4 = 15x^4$;   when $x = 2$, $\dfrac{dy}{dx} = 15(2)^4 = 240$.

  b. Here, $n = 0$, so $\dfrac{dy}{dx} = 0$ at all values of x.

  c. $n = -1$; $\dfrac{dy}{dx} = (-1)(2)x^{-2} = \dfrac{-2}{x^2} = -\dfrac{1}{2}$ at $x = 2$.

It is possible, by various means, to determine the derivatives of a variety of functions, several of which are presented without proof in Table 9.1. It is interesting to note that the functions e^x and $\ln x$ give particularly simple derivatives, e^x and $1/x$. This explains, at least in part, why expressions involving natural logarithms or exponentials to the base e appear so frequently in mathematical equations describing physical laws.

Table 9.1

Function	Derivative
1. $y = a$	1. $dy/dx = 0$
2. $y = ax^n$	2. $dy/dx = nax^{(n-1)}$
3. $y = \ln x$	3. $dy/dx = 1/x$
4. $y = \log_{10} x$	4. $dy/dx = 1/(2.30\,x)$
5. $y = e^x$	5. $dy/dx = e^x$
6. $y = a^x$	6. $dy/dx = a^x \ln a$
7. $y = \sin x$	7. $dy/dx = \cos x$
8. $y = \cos x$	8. $dy/dx = -\sin x$

EXERCISES

Determine dy/dx for the following functions:

1. $-4x^2$, at $x = 1$
2. $x^3/8$, at $x = 3$
3. e^x, at $x = 1$
4. $\ln x$, at $x = 2$
5. $\log_{10} x$, at $x = 1.20$
6. $\sin x$, at $x = 30°$
7. $\cos x$, at $x = 30°$
8. 3^x, at $x = -2$

9.3 GENERAL RULES FOR DIFFERENTIATION

Using Table 9.1, we can find the derivatives of simple functions such as ax^n, $\ln x$, e^x and $\sin x$. At this stage, however, we cannot take the derivative of such expressions as:

$y = 2x^2 + 3x$ (sum of two functions)

$y = xe^x$ (product of two functions)

$y = \dfrac{\ln x}{x^2}$ (quotient of two functions)

$y = (e^x + 1)^{1/2}$ (function of a function)

In this section, we shall present general rules for handling each of these cases.

Sum of Two Functions. The rule here is an obvious one; **the derivative of the sum of two functions is the sum of the derivatives of the individual functions.** That is, if:

$$y = u + v \tag{9.12}$$

where u and v are functions of x, then:

$$\frac{dy}{dx} = \frac{du}{dx} + \frac{dv}{dx} \tag{9.13}$$

This rule applies to the difference between two functions. That is, if:

$$y = u - v; \qquad \frac{dy}{dx} = \frac{du}{dx} - \frac{dv}{dx} \tag{9.14}$$

It can also be extended to any number of functions. For example, if:

$$y = u + v + w; \qquad \frac{dy}{dx} = \frac{du}{dx} + \frac{dv}{dx} + \frac{dw}{dx} \tag{9.15}$$

Example 9.3 If $y = 2x^2 + 3x$, find dy/dx.

Solution

$$\frac{dy}{dx} = \frac{d(2x^2)}{dx} + \frac{d(3x)}{dx} = 4x + 3$$

Product of Two Functions. If y is expressed as the product of two functions of x, u and v, i.e., if:

$$y = uv$$

then:

$$\frac{dy}{dx} = u\frac{dv}{dx} + v\frac{du}{dx} \tag{9.16}$$

Example 9.4 Find dy/dx if:
a. $y = xe^x$
b. $y = x^2 \ln x$

Solution
a. Let $u = x$, $v = e^x$
$$\frac{dy}{dx} = x\frac{d(e^x)}{dx} + e^x\frac{d(x)}{dx}$$

$$= xe^x + e^x (1) = e^x (x + 1)$$

b. Let $u = x^2, v = \ln x$

$$\frac{dy}{dx} = x^2 \frac{d (\ln x)}{dx} + \ln x \frac{d (x^2)}{dx}$$

$$= x^2 \left(\frac{1}{x}\right) + \ln x (2x) = x + 2x \ln x$$

An important corollary to this rule applies to the situation in which one of the "functions" of x is actually a constant.

$$\text{if } y = av \qquad (a = \text{constant}, v = \text{function of } x)$$

$$\text{then } \frac{dy}{dx} = a \frac{dv}{dx} + v \frac{da}{dx}$$

But $\frac{da}{dx}$ must be 0 (Table 9.1). Hence:

$$\frac{dy}{dx} = a \frac{dv}{dx} \qquad (9.17)$$

Applying this rule to the function:

$$y = 5 \sin x$$

we have:

$$\frac{dy}{dx} = 5 \frac{d \sin x}{dx} = 5 \cos x$$

Quotient of Two Functions. If $y = u/v$, we can always find dy/dx by applying the multiplication rule, taking the two functions to be u and $1/v$ (See Exercise 2). In many cases, it is simpler to use the rule:

$$\frac{dy}{dx} = \frac{v \dfrac{du}{dx} - u \dfrac{dv}{dx}}{v^2} \qquad (9.18)$$

Example 9.5 Find dy/dx if:

a. $y = \dfrac{e^x}{x}$

b. $y = \dfrac{\ln x}{x^2}$

Solution

a. Let $u = e^x$, $v = x$:

$$\frac{dy}{dx} = \frac{x\dfrac{d(e^x)}{dx} - e^x\dfrac{d(x)}{dx}}{x^2} = \frac{xe^x - e^x}{x^2}$$

b. Here, $u = \ln x$, $v = x^2$:

$$\frac{dy}{dx} = \frac{x^2\dfrac{d(\ln x)}{dx} - \ln x\dfrac{d(x^2)}{dx}}{x^4} = \frac{\dfrac{x^2}{x} - 2x\ln x}{x^4}$$

$$= \frac{1 - 2\ln x}{x^3}$$

Compound Functions. The Chain Rule. Functions such as:

$$y = (x^2 - 1)^{-1} \quad \text{or} \quad y = (e^x + 1)^{1/2}$$

are sometimes referred to as compound functions because they can be thought of as functions of a variable, u, which is itself a function of x. Thus:

$$y = (x^2 - 1)^{-1} = u^{-1}, \text{where } u = x^2 - 1$$

$$y = (e^x + 1)^{1/2} = u^{1/2}, \text{where } u = e^x + 1$$

Derivatives of compound functions can be obtained by applying the **chain rule:**

$$\frac{dy}{dx} = \frac{dy}{du}\frac{du}{dx} \qquad (9.19)$$

To illustrate how this rule is used, let us consider how one would find the derivative of $(x^2 - 1)^{-1}$:

$$y = u^{-1}, \text{where } u = x^2 - 1$$

$$\frac{dy}{du} = -u^{-2} = \frac{-1}{(x^2 - 1)^2}$$

$$\frac{du}{dx} = 2x$$

Hence:

$$\frac{dy}{dx} = \frac{-2x}{(x^2 - 1)^2}$$

Example 9.6   Find dy/dx if:

  a. $y = (e^x + 1)^{1/2}$
  b. $y = (x^2 - 2)^3$

Solution

  a. Let $u = e^x + 1$.
      Then $y = u^{1/2}$

$$\frac{dy}{du} = \frac{1}{2} u^{-1/2} = \frac{1}{2(e^x + 1)^{1/2}}$$

$$\frac{du}{dx} = e^x$$

  Hence: $$\frac{dy}{dx} = \frac{dy}{du}\frac{du}{dx} = \frac{e^x}{2(e^x + 1)^{1/2}}$$

  b. In principle, we could expand $(x^2 - 2)^3$ and take the derivative of each term. It is simpler to take:

$$u = x^2 - 2; \qquad y = u^3$$

$$\frac{dy}{du} = 3u^2 = 3(x^2 - 2)^2$$

$$\frac{du}{dx} = 2x$$

  Hence: $$\frac{dy}{dx} = 6x(x^2 - 2)^2$$

EXERCISES

1. Using the method employed in Section 9.2 to show that $\dfrac{d(ax^2)}{dx} = 2ax$, prove that:

  a. $\dfrac{d(u + v)}{dx} = \dfrac{du}{dx} + \dfrac{dv}{dx}$

  b. $\dfrac{d(uv)}{dx} = u\dfrac{dv}{dx} + v\dfrac{du}{dx}$

2. Use the rule for multiplication given in 9.16 to prove the rule for division given in 9.18.

3. Find dy/dx for each of the following functions:

a. $y = x^5 + 3x^2 - 2x$ d. $y = (\ln x)^3$

b. $y = x^2 \sin x - \cos x$ e. $y = \dfrac{\sin x}{\cos x}$

c. $y = 3e^x \cos x$ f. $y = \dfrac{(e^x - 1)^2}{e^x + 1}$

9.4 MAXIMA AND MINIMA

Certain functions, of which the sine curve shown in Figure 9.2a is an example, go through one or more maxima or minima. Point A in Figure 9.2a represents a relative maximum, inasmuch as the value of y (i.e., $\sin x$) at that point is greater than at values of x immediately preceding or immediately following it. By the same token, point B represents a relative minimum.

The principles of differential calculus can be used to locate the positions of relative maxima and minima. In Figure 9.2a, we have drawn dotted lines to represent tangents to the curve at points A and B. These tangents are parallel to the x axis; their slopes are zero. Since the derivative, dy/dx, can be taken as the slope of the tangent to the curve, it follows that dy/dx *must be zero at a relative maximum or minimum.*

However, it turns out that the condition that $dy/dx = 0$ is not sufficient in itself to prove the existence of a relative maximum or minimum. Consider the curve shown in Figure 9.2b. Point C is clearly neither a maximum nor a minimum. Yet the slope of the tangent, and hence dy/dx, is 0 (points of this type are referred to as *points of inflection*). How, then, are

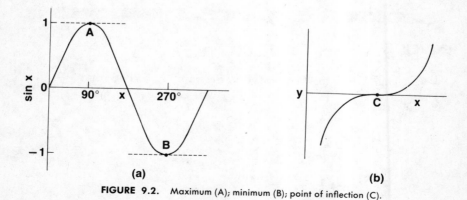

FIGURE 9.2. Maximum (A); minimum (B); point of inflection (C).

we to tell whether we are dealing with a relative maximum, a relative minimum, or a point of inflection when, at a particular value of x, we find that $dy/dx = 0$?

This question can be answered by considering how dy/dx changes in the vicinity of points A, B, and C. From Figure 9.2, we note that to the left of the maximum at A, dy/dx is a positive quantity (y increases as x increases); to the right of A, dy/dx has a negative sign (y decreases as x increases). At the relative minimum at point B, we find the reverse situation. To the left of B, dy/dx is negative (falling curve); to the right of B, it is positive (rising curve). In contrast to this behavior, we find that, on both sides of point C, dy/dx has the same sign. (In this case, the slope is positive on both sides of the point of inflection; it is also possible to have a point of inflection on a falling curve, in which case dy/dx would be a negative quantity on both sides of the point.)

These ideas are summarized in Table 9.2.

Table 9.2 Conditions that $y = f(x)$ Go Through a Maximum, Minimum, or Point of Inflection at $x = a$

	Value of dy/dx		
	$x < a$	$x = a$	$x > a$
Maximum	+	0	−
Minimum	−	0	+
Point of inflection	+	0	+
Point of inflection	−	0	−

From Table 9.2, we deduce that to locate a relative maximum, relative minimum, or point of inflection in the curve corresponding to the function $y = f(x)$, we:

1. Obtain an expression for dy/dx.
2. Set $dy/dx = 0$ and solve for x.
3. Use Table 9.2 to decide whether y goes through a maximum, minimum, or point of inflection at this value of x.

Example 9.7 For the function $y = x^3 - 3x$, find the points at which $dy/dx = 0$, and determine whether they represent relative maxima, relative minima, or points of inflection.

Solution Taking the derivative, we obtain:

$$\frac{dy}{dx} = 3x^2 - 3$$

Setting $\dfrac{dy}{dx} = 0$ and solving:

$$3x^2 = 3; \qquad x^2 = 1; \qquad x = \pm 1$$

To decide whether $x = 1$ represents a relative maximum, minimum, or point of inflection, let us find dy/dx when x is slightly less than 1, let us say 0.90, and slightly greater than 1, e.g., 1.10:

When $x = 0.90, \dfrac{dy}{dx} = 3(0.90)^2 - 3 = 2.4 - 3 = -0.6$

When $x = 1.10, \dfrac{dy}{dx} = 3(1.10)^2 - 3 = 3.6 - 3 = +0.6$

We see that when $x < 1$, $dy/dx < 0$, while when $x > 1$, $dy/dx > 0$. Clearly, the point $x = 1$ represents a minimum (see Table 9.2). By precisely the same line of reasoning, we conclude that at the point $x = -1$, we have a maximum.

$x = -1.10; \qquad dy/dx = 3.6 - 3 = 0.6; \qquad dy/dx > 0$

$x = -0.90; \qquad dy/dx = 2.4 - 3 = -0.6; \qquad dy/dx < 0$

EXERCISES

Find relative maxima, minima, or points of inflection for the following functions:

1. $y = x^2 + 6x$
2. $y = x^2 - 6x$
3. $y = x^3 - 4x^2$

4. $y = x^3 - 6x^2 + 3x$
5. $y = x \ln x$
6. $y = \sin x$ ($x = 0$ to $x = 180°$)

9.5 DIFFERENTIALS

Instead of thinking of the derivative, dy/dx, as a single, indivisible quantity, it is often convenient to regard it as the ratio of two separate quantities, dy and dx, which we refer to as differentials. When we apply differential calculus to the function:

$$y = x^2$$

we can write: $\dfrac{dy}{dx} = 2x$

or, equally well: $dy = 2x\, dx$

As these equations imply, differentials may be treated algebraically like any other quantities. We can, for example, multiply both sides of an equation by dx. This explains why the formulas we have written for derivatives apply equally well to differentials (Table 9.3).

Table 9.3   Differentials

1. $d\,a = 0$	6. $d\cos x = -\sin x\, dx$
2. $d(ax^n) = anx^{(n-1)}\,dx$	7. $d(u + v) = du + dv$
3. $d\ln x = dx/x$	8. $d(uv) = u\,dv + v\,du$
4. $d\,e^x = e^x\,dx$	9. $d(u/v) = \dfrac{v\,du - u\,dv}{v^2}$
5. $d\sin x = \cos x\,dx$	10. $dy = \left(\dfrac{dy}{du}\right)du$

In each case, the rule for differentials is obtained by multiplying both sides of the corresponding equation for the derivative (see Table 9.1) by dx.

The rules given in Table 9.3 offer a simple way to obtain differentials (or derivatives) of *implicit functions*, i.e., relations in which y is not given explicitly as a function of x. Consider, for example, the equation:

$$xy^2 = 6$$

Applying rule 8 of Table 9.3, with $x = u$ and $y^2 = v$, we have:

$$d(xy^2) = x(2y\,dy) + y^2\,dx = 0$$

To obtain an expression for dy, we rearrange to obtain:

$$dy = \frac{-y^2}{2xy}\,dx = \frac{-y}{2x}\,dx$$

If we wish to obtain an expression for dy/dx, we divide both sides of this equation by dx:

$$\frac{dy}{dx} = \frac{-y}{2x}$$

Example 9.8   If $xe^y = x^2 - 1$, obtain expressions for dy and dy/dx

Solution $d(xe^y) = d(x^2 - 1)$

The derivative of xe^y may be found by considering it to be a product:

$$d(xe^y) = x\,d(e^y) + e^y\,dx$$

$$= xe^y\,dy + e^y\,dx$$

The derivative of $x^2 - 1$ is readily found:

$$d(x^2 - 1) = 2x\,dx$$

Equating the two derivatives that we have evaluated:

$$xe^y\,dy + e^y\,dx = 2x\,dx$$

Solving for dy: $dy = \dfrac{(2x - e^y)\,dx}{xe^y}$

Hence, $\dfrac{dy}{dx} = \dfrac{2x - e^y}{xe^y}$

As we have seen, the differential dy is related to dx by the *exact* equation:

$$dy = \left(\frac{dy}{dx}\right) dx \tag{9.20}$$

At small values of Δx, we can write the *approximate* equation:

$$\Delta y \approx \left(\frac{dy}{dx}\right) \Delta x \tag{9.21}$$

where Δy is the change in y brought about by the small change in x, Δx.

Equation 9.21 is an extremely useful one for obtaining a rapid estimate of the effect upon y of a small change in the independent variable, x. To illustrate its use, suppose that for the function:

$$y = x^4$$

we wish to know Δy when x changes from 10.0 to 10.1. Applying Equation 9.21:

$$\Delta y \approx \left(\frac{dy}{dx}\right) \Delta x = 4x^3\,\Delta x = 4(10.0)^3(0.1) = 400$$

We could, of course, solve this problem the hard way by finding y when $x = 10.1$:

$$(10.1)^4 = 10,406.0401$$

$$(10.0)^4 = 10,000.0000$$

$$\Delta y = 406.0401$$

Comparing our approximate value for Δy, 400, to the exact value, 406.0401, we see that the approximate value is in error by about 6 parts in 400, or 1.5 per cent (Notice that the value of y itself, using Equation 9.21 to obtain Δy, would be 10,400, which is in error by only 6 parts in 10,000, or 0.06 per cent!)

The smaller Δx is, the less will be the error in the value of Δy that we calculate using Equation 9.21. If Δx in the example just cited were 0.01 instead of 0.10, the percent error in Δy would be considerably less than 1.5 per cent. Indeed, one can write the exact equation:

$$\underset{\Delta x \to 0}{\text{Limit}} \ \Delta y = \left(\frac{dy}{dx}\right) \Delta x \qquad (9.22)$$

Example 9.9 If $y = \ln x$, what is Δy when x changes from 1.000 to 1.001?

Solution

$$\Delta y \approx \left(\frac{dy}{dx}\right) \Delta x = \frac{1}{x}(\Delta x) = \frac{0.001}{1} = 0.001$$

Or, since $\ln 1.000 = 0$, we could write:

$$\ln 1.001 = 0.001$$

In general:

$$\ln (1 + a) \approx a, \text{ when } a \text{ is very small.}$$

This gives us a convenient way of estimating natural (or base 10) logs of numbers close to unity.

Equation 9.21 can be used to estimate the effect of a known experimental error (Δx) in one variable upon the calculated value of a dependent variable (Δy). Example 9.10 illustrates how this is done.

Example 9.10 The density of a liquid is determined by measuring the mass and volume of a sample and using the equation:

$$\rho = \frac{m}{V}$$

In a particular experiment, a student finds m to be 6.00 g and $V = 8.0$ ml. What is the error in ρ (i.e., $\Delta\rho$), if:

    a. V is exactly 8.0 ml, but m is in error by 0.010 g (i.e., $\Delta m = 0.010$ g)?

    b. m is exactly 6.00 g, but V is in error by 0.10 ml (i.e., $\Delta V = 0.10$ ml)?

Solution

    a. $\Delta\rho = \dfrac{d\rho}{dm} \Delta m$

        Taking V to be a constant: $\dfrac{d\rho}{dm} = \dfrac{1}{V}$

        Hence: $\Delta\rho = \dfrac{\Delta m}{V} = \dfrac{0.010 \text{ g}}{8.0 \text{ ml}} = 0.0013$ g/ml

    b. $\Delta\rho = \dfrac{d\rho}{dV} \Delta V = \dfrac{-m}{V^2} \Delta V = \dfrac{-6.00 \text{ g}}{64.0 \text{ (ml)}^2} \times 0.10$ ml

$$= -0.0094 \text{ g/ml}$$

In other words, if the measured mass were 0.01 g greater than the true mass, the calculated density would be 0.0013 g/ml larger than the true density. If the measured volume were too high by 0.10 ml, the calculated density would be 0.0094 g/ml less than the true density.

EXERCISES

    1. Find expressions for dy/dx when:

        a. $y^2 x^2 = 8$ c. $x \ln y = 3$

        b. $xy^{1/2} = y^2 - 4$ d. $y^2 e^x = 2$

    2. A spherical balloon of radius 12.00 cm is expanded until its radius becomes 12.01 cm. By how much does its volume increase?

3. Following Example 9.9, calculate and compare to the values found in tables:
   a. log 1.010
   b. log 0.990
   c. log 2.030, given log 2.000 = 0.3010
4. The solubility product of silver chloride is given by the expression:

$$K_{sp} = (\text{conc. Ag}^+)(\text{conc. Cl}^-)$$

In a particular experiment it is found that the concentration of $Ag^+ = 0.10$ and the concentration of $Cl^- = 1.6 \times 10^{-9}$.
   a. Calculate K_{sp}.
   b. What is the error in the calculated value of K_{sp} caused by a 10 per cent error in the concentration of Ag^+, assuming the concentration of Cl^- is correct?
   c. What is the error in the calculated value of K_{sp} caused by a 20 per cent error in the concentration of Cl^-, assuming no error in the concentration of Ag^+?
5. Show that if the mass of a liquid sample is known exactly, the percentage of error in the density, $100 \, \Delta\rho/\rho$, is approximately equal in magnitude to the percentage of error in the volume, $100 \, \Delta V/V$.

PROBLEMS

9.1 Plot the data given in Section 9.1 for the concentration of NO_2 as a function of time. By drawing tangents to the curve, estimate as accurately as you can the rate of reaction, d conc. NO_2/dt, at $t = 0$, $t = 1$, $t = 2$, and $t = 3$.

9.2 The relationship between temperature in Fahrenheit and Centigrade degrees is:

$$°F = 1.8°C + 32°$$

   a. Calculate $\Delta°F$ and $\dfrac{\Delta°F}{\Delta°C}$ when $\Delta°C = 1$.

   b. Calculate $\Delta°F$ and $\dfrac{\Delta°F}{\Delta°C}$ when $\Delta°C = 0.1$.

   c. Calculate $\dfrac{d°F}{d°C}$. Explain why the derivative is equal to $\Delta°F/\Delta°C$.

9.3   The volume, V, of one gram of water is given as a function of Centigrade temperature, t, by the equation:

$$V = V_0 (1 - 4.5 \times 10^{-5}t + 6.7 \times 10^{-6}t^2 - 1.9 \times 10^{-8}t^3)$$

where V_0 is the volume at $0°C$. Calculate dV/dt, in terms of V_0, at:
   a. $0°C$ c. $25°C$
   b. $4°C$ d. $100°C$

9.4   In any aqueous solution at $25°C$, the concentrations of H^+ and OH^- are related by the expression:

$$\text{conc. } H^+ = \frac{1.0 \times 10^{-14}}{\text{conc. } OH^-}$$

   a. Find a general expression for $d(\text{conc. } H^+)/d(\text{conc. } OH^-)$.
   b. Evaluate this derivative at conc. $OH^- = 1.0$, 1.0×10^{-7}, 1.0×10^{-14}.

9.5   The standard free energy change, $\Delta G°$, for a reaction is given by the expression:

$$\Delta G° = \Delta H° - T\Delta S°$$

where $\Delta H°$ and $\Delta S°$ are the standard enthalpy and entropy changes respectively and T is the absolute temperature.
   a. Evaluate $d\Delta G°/dT$.
   b. Show that $\Delta G°$ will increase with temperature if $\Delta G° > \Delta H°$.

9.6   The reduction potential, E, for the half reaction:

$$2H^+ + 2e^- \rightarrow H_2 (g, 1 \text{ atm})$$

is given by the expression:

$$E = 0.059 \log (\text{conc. } H^+)$$

Calculate $dE/d(\text{conc. } H^+)$ at:
   a. conc. $H^+ = 1$
   b. conc. $H^+ = 10^{-7}$

9.7   The equilibrium constant for a reaction, K_p, is related to the standard free energy change, $\Delta G°$, by the relation:

$$\ln K_p = \frac{-\Delta G°}{RT}, \text{ where } R \text{ is the gas law constant.}$$

The variation of $\Delta G°$ with temperature is given in Problem 9.5. Show that

$$\frac{d \ln K_p}{dT} = \frac{\Delta H°}{RT^2}$$

(Hint: use the rule given in Section 9.3 for differentiating quotients.)

9.8 For the reaction:

$$CH_3CHO(g) \rightarrow CO(g) + CH_4(g)$$

the concentration of $CH_3CHO(g)$ is found to vary with time (t) according to the relation:

$$y = \frac{1}{kt + c} \qquad (k \text{ and } c \text{ are constants})$$

Obtain an expression for the rate of reaction, $-dy/dt$, and show that it is equal to ky^2 (start by letting $u = kt + c$).

9.9 The basic law of radioactive decay is:

$$A = A_0 e^{-kt}$$

where A is the activity at time t, A_0 the activity at time zero, and k is a constant. Show that $-dA/dt = kA$.

9.10 For a certain reaction, it is found that the concentration, y, is given as a function of time, t, by the relation:

$$\frac{1}{y^2} - \frac{1}{y_0^2} = 2kt$$

where y_0 is the concentration at $t = 0$. Show that for this reaction, $-dy/dt = ky^3$ (use the method illustrated in Example 9.8).

9.11 Using the equation given in Problem 9.3, find the temperature at which $dV/dt = 0$ for water. Does this represent a maximum, minimum, or point of inflection? Draw a sketch of the curve of V vs t for water.

9.12 If the relationship between two experimental quantities, y and x, is:

$$y = ax^2$$

show that the percentage of error in y $(100\ \Delta y/y)$ is approximately twice as great as the percentage of error in x.

9.13 The vapor pressure of a liquid at a given temperature can be calculated from the equation:

$$\log P = -\frac{\Delta H_{vap}}{4.58\ T} + \text{constant}$$

    a. Estimate the error in $\log P$ at $T = 300°K$ if ΔH_{vap} is in error by 100 cal.

    b. Estimate the percentage of error in P under the conditions in part (a).

CHAPTER 10

INTEGRAL CALCULUS

Differential calculus, discussed in Chapter 9, enables us to find dy/dx, given y as a function of x. Integral calculus addresses itself to the reverse problem. Given an expression for dy/dx, find the functional relationship between y and x. A knowledge of the techniques of integration is particularly useful in the fields of chemical kinetics and chemical thermodynamics.

10.1 THE INDEFINITE INTEGRAL

To illustrate the type of problem that can be solved by integral calculus, let us suppose that we are given the relation:

$$\frac{dy}{dx} = x$$

How can we find the functional relationship between y and x? That is, what must be the relationship between y and x in order for the derivative, dy/dx, to be the quantity x?

We can answer this question by recalling from Chapter 9 that the derivative of x^2 is $2x$. That is:

$$\text{if } y = x^2, \quad \text{then } \frac{dy}{dx} = 2x$$

Now, we want the function whose derivative is x rather than $2x$. A moment's reflection should convince you that the required function is $x^2/2$.

143

That is:

$$\text{if } y = x^2/2, \quad \text{then } \frac{dy}{dx} = \frac{2x}{2} = x$$

We deduce then that a function whose derivative is x must be:

$$y = x^2/2$$

The process which we have just described is basic to integral calculus. What we have done is to integrate the differential equation:

$$dy = x\,dx$$

to obtain the function $y = x^2/2$. In mathematical symbolism, the operation of integration is indicated by the symbol $\int$. We write:

$$\int dy = \int x\,dx$$

to indicate that we are required to find the finite quantities whose differentials are dy and dx. Realizing that dy is simply the differential of y, we might rewrite this equation as:

$$y = \int x\,dx$$

with the understanding that we are to find the function whose differential is $x\,dx$. In general:

$$\text{if} \qquad dy = F(x)\,dx$$

$$\text{then} \qquad \int dy = \int F(x)\,dx$$

$$\text{or} \qquad y = \int F(x)\,dx \qquad\qquad (10.1)$$

Let us now ask the question: Is the relation $y = x^2/2$ the *only* function which will satisfy the requirement that $dy/dx = x$? As a matter of fact, it is not; the functions:

$$y = x^2/2 + 6; \qquad y = x^2/2 - 8; \qquad y = x^2/2 + 3$$

would all yield on differentiation:

$$dy/dx = x$$

Indeed, there is a whole family of functions:

$$y = \frac{x^2}{2} + C \qquad (C = \text{any constant})$$

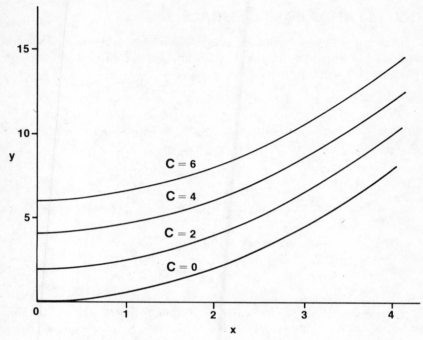

FIGURE 10.1. Plot of function: $y = \dfrac{x^2}{2} + C$.

for which $dy/dx = x$. A few of these equations are plotted in Figure 10.1. You will note that at any given value of x, the slopes of the tangents to these curves and hence the derivatives are identical.

This discussion illustrates an important distinction between the processes of differentiation and integration. Given a particular function, there is one and only one expression for the derivative, dy/dx. That is:

$$\text{if } y = f(x), \quad \text{then } \frac{dy}{dx} = F(x)$$

where $F(x)$ refers to a specific function of x. On the other hand, corresponding to a given derivative, dy/dx, there is a family of functions which differ from each other by a constant term, C.

If: $$\frac{dy}{dx} = F(x), \quad \text{then } y = \int F(x)\, dx = f(x) + C \qquad (10.2)$$

The constant appearing in Equation 10.2 is often referred to as the **constant of integration;** later in this chapter, we shall show how one might evaluate it in a particular case.

10.2 STANDARD INTEGRALS

In Section 10.1, we obtained the integral of xdx by, in essence, reversing the process of differentiation. Many simple differentials are susceptible to this same approach. For example, to find:

$$\int x^2 dx$$

we recall that differentiation of x^3 gives $3x^2 dx$. It follows that $x^2 dx$ must be the differential of $x^3/3$. In other words:

$$\int x^2 dx = x^3/3 + C$$

Again, to find: $\quad\int \dfrac{dx}{x}$

We recall that: $\quad d\ln x = \dfrac{dx}{x}$

Hence: $\quad\int \dfrac{dx}{x} = \ln x + C$

In many cases, it is not nearly as obvious what family of functions will lead to a given differential. A few of the more common integrals are listed in Table 10.1 Much more extensive tables of integrals are available in handbooks, such as *Handbook of Chemistry and Physics*, Chemical Rubber Publishing Co., Cleveland, Ohio, 1965, and N. A. Lange: *Handbook of Chemistry*. McGraw-Hill, New York, 1967.

Example 10.1

  a. Given $dy/dx = x^{-2}$, find y as a function of x.
  b. Integrate: $\dfrac{x\,dx}{2x + 3}$

Solution

  a. $\dfrac{dy}{dx} = x^{-2}; \qquad dy = x^{-2}dx; \qquad y = \int x^{-2}dx$

  Using relation (2) in Table 10.1, with $n = -2$:

  $$y = \frac{x^{-2+1}}{-2+1} + C = -x^{-1} + C = -\frac{1}{x} + C$$

  b. We use relation (11) in Table 10.1, noting that $a = 2$ and $b = 3$:

  $$\int \frac{x\,dx}{2x + 3} = \frac{x}{2} - \frac{3}{4}\ln(2x + 3) + C$$

Table 10.1   Standard Integrals

1. $\int dx = x + C$

2. $\int x^n dx = \dfrac{x^{(n+1)}}{n + 1} + C \qquad (n \neq -1)$

3. $\int dx/x = \ln x + C$

4. $\int e^x dx = e^x + C$

5. $\int a^x dx = \dfrac{a^x}{\ln a} + C$

6. $\int \ln x \, dx = x \ln x - x + C$

7. $\int \sin x \, dx = -\cos x + C$

8. $\int \cos x \, dx = \sin x + C$

9. $\int (ax + b)^n dx = \dfrac{(ax + b)^{n+1}}{a(n + 1)} + C \qquad (n \neq -1; \text{compare relation [2]})$

10. $\int dx/(ax + b) = \dfrac{\ln(ax + b)}{a} + C \qquad (\text{compare relation [6]})$

11. $\int \dfrac{x \, dx}{ax + b} = \dfrac{x}{a} - \dfrac{b \ln(ax + b)}{a^2} + C$

12. $\int \dfrac{x \, dx}{(ax + b)^2} = \dfrac{b}{a^2(ax + b)} + \dfrac{1}{a^2} \ln(ax + b) + C$

13. $\int \dfrac{x^2 \, dx}{ax + b} = \dfrac{1}{a^3} \left[\dfrac{(ax + b)^2}{2} - 2b(ax + b) + b^2 \ln(ax + b) \right] + C$

In each relation listed in Table 10.1, a constant of integration is included for reasons pointed out in Section 10.1. (The constant C is sometimes omitted in tables of integrals appearing in handbooks; these same sources often use "log" to refer to a natural logarithm!) When we apply these formulas to physical problems, it is often possible to deduce the numerical value of C. In particular, C can be calculated if y is known at one value of x (Example 10.2).

Example 10.2. Given that $dy/dx = \ln x$ and $y = 0$ when $x = 1$, find $y = f(x)$.

Solution The general function is found by applying relation (6) in Table 10.1.

$$\int dy = \int \ln x \, dx$$

$$y = x \ln x - x + C$$

To evaluate C, we note that $y = 0$ when $x = 1$. Hence:

$$0 = 1 (\ln 1) - 1 + C$$

$$0 = 0 - 1 + C$$

$$C = 1$$

The function must then be: $y = x \ln x - x + 1$

EXERCISES

1. Obtain general expressions for the following integrals:

   a. $\int x^3 dx$ c. $\int 5^x dx$

   b. $\int e^x dx$ d. $\dfrac{x \, dx}{(x + 3)^2}$

2. Obtain the particular function for $1a - 1d$ if:
   a. $y = 0$ when $x = 1$
   b. $y = 1$ when $x = 1$

10.3   RULES FOR INTEGRATION

Frequently, we are required to integrate expressions that are not included in Table 10.1, or, for that matter, in more extensive tables. In this regard, there are two very simple rules that we will find useful:

1. $\int a F(x) \, dx = a \int F(x) \, dx$ (10.3)

   e.g., $\int 5 x^2 dx = 5 \int x^2 dx = \dfrac{5}{3} x^3 + C$

2. $\int [F(x) + F'(x)] \, dx = \int F(x) \, dx + \int F'(x) \, dx$ (10.4)

   e.g., $\int (x + e^x) \, dx = \int x \, dx + \int e^x dx = \dfrac{x^2}{2} + e^x + C$

(Note that it is only necessary to include one constant of integration; the sum of two constants is itself a constant.)

Example 10.3   Find the functions for which:
a. $dy/dx = 6e^x$
b. $dy/dx = (x^2 + \ln x)$

Solution

$$a. \quad y = \int 6e^x \, dx = 6 \int e^x \, dx = 6e^x + C$$

$$b. \quad y = \int (x^2 + \ln x) \, dx$$
$$= \int x^2 \, dx + \int \ln x \, dx$$

Using relations (2) and (6) of Table 10.1:

$$y = \frac{x^3}{3} + x \ln x - x + C$$

For more complex cases, several techniques are available. We shall discuss only one of these; others may be found in elementary calculus books.

Substitution of Variable. We sometimes find that an expression which cannot be integrated directly is readily handled by making a simple change in variables. Suppose, for example, we are required to find:

$$y = \int e^{2x} \, dx$$

This integral does not exactly match any of those listed in Table 10.1. However, if we let:

$$u = 2x$$

it becomes possible to arrive at an expression, involving terms in u and du, which can be integrated.

To obtain such an expression, we substitute $u = 2x$ to get:

$$y = \int e^u \, dx$$

Before we can integrate this expression, we must convert the differential dx to its equivalent in terms of du. To do this, we note that:

$$u = 2x$$
$$du = 2dx; \qquad dx = du/2$$

Making this substitution:

$$y = \int \frac{e^u}{2} \, du = \frac{1}{2} \int e^u \, du = \frac{e^u}{2} + C$$

To get y in terms of x, we replace u in the final expression by $2x$ and write:

$$y = \frac{e^{2x}}{2} + C$$

The general procedure which we follow in applying this method may be summarized as follows. Given:

$$y = \int F(x)\,dx$$

choose a new variable, u, so as to convert the integral to a form:

$$y = \int F(u)\,du$$

which can be integrated using one of the relations listed in Table 10.1 or in a more extensive table of integrals. The trick, obviously, is to decide what the new variable, u, should be to give us an integrable expression. Sometimes, as in the example just worked, the choice of u is rather obvious. In other cases, it may not be nearly so easy to decide what function of x we should let u represent. No general rule can be given; the proper function sometimes has to be found by trial and error.

Example 10.4   Obtain expressions for:

a. $\int 2x\,(x^2 + 3)^{1/2}\,dx$

b. $\dfrac{\ln x}{x}\,dx$

Solution

a. A reasonable choice for u would seem to be:

$$u = x^2 + 3$$

Then: $du = 2x\,dx$
Our integral now becomes:

$$\int 2x\,(x^2 + 3)^{1/2}\,dx = \int u^{1/2}\,du$$

$$= \frac{2}{3}\,u^{3/2} + C$$

$$= \frac{2}{3}\,(x^2 + 3)^{3/2} + C$$

b. Let us try the substitution:

$$u = \ln x$$

Then: $du = \dfrac{1}{x} dx$

and we obtain: $\displaystyle\int \dfrac{\ln x}{x} dx = \int u\, du = \dfrac{u^2}{2} + C = \dfrac{(\ln x)^2}{2} + C$

In this case, the substitution $u = ln\,x$ worked very nicely. However, had the integral been:

$$\int \dfrac{\ln x}{x^2} dx$$

the same substitution would have been ineffective (try it!).

EXERCISES

Evaluate the following integrals:

1. $\displaystyle\int 3\,e^x dx$

2. $\displaystyle\int e^{6x} dx$

3. $\displaystyle\int (e^x + \ln x)\, dx$

4. $\displaystyle\int \dfrac{dx}{(1 - 2x)^3}$

5. $\displaystyle\int \ln (1 - x)\, dx$

6. $\displaystyle\int (x + 2)^{1/2} dx$

7. $\displaystyle\int \dfrac{e^x}{e^x + 1}\, dx$

10.4 THE DEFINITE INTEGRAL

Up to this point, we have directed our attention to one specific problem: Given an expression for dy/dx, find the corresponding function, $y = f(x)$. Frequently, the problem that confronts us in integral calculus is a somewhat different one: Given an expression for dy/dx, find the difference between the values of y at two different values of x.

To illustrate the problem, suppose that:

$$\frac{dy}{dx} = x$$

and that we are required to find Δy when x changes from an initial value of 2 to a final value of 3. We might start by obtaining the general functional relationship between y and x.

$$y = \frac{x^2}{2} + C$$

Now, when $x = 3$: $y_{final} = \frac{9}{2} + C$

when $x = 2$: $y_{initial} = \frac{4}{2} + C$

Hence: $\Delta y = y_{final} - y_{initial} = \frac{9}{2} - \frac{4}{2} = \frac{5}{2}$

It should be clear from this illustration that Δy represents the difference between the integral evaluated at $x = 3$ and $x = 2$. In the language of integral calculus, we write:

$$\Delta y = \int_2^3 x\, dx$$

The quantity on the right of this equation is referred to as a **definite integral**, which is to be evaluated by subtracting its value at the **lower limit** ($x = 2$) from that at the **upper limit** ($x = 3$). We sometimes indicate this by writing:

$$\Delta y = \left. \frac{x^2}{2} \right|_2^3$$

and it is understood that we are to subtract the value of $x^2/2$ at the lower limit ($x^2/2 = 4/2$) from that at the upper limit ($x^2/2 = 9/2$). Note that in evaluating a definite integral, *it is not necessary to include the constant of integration, C, since it disappears when the subtraction is carried out.*

These observations can be expressed in general terms in the following statements:

1. If $dy = F(x)\, dx$, then the change in y which results when x changes from an initial value, a, to a final value, b, is:

$$\Delta y = \int_a^b F(x)\, dx \tag{10.5}$$

2. If the indefinite integral, $\int F(x)\, dx = f(x) + C$, then the expression for the definite integral is:

$$\int_a^b F(x)\, dx = f(x) \left. \right|_a^b = f(x = b) - f(x = a) \tag{10.6}$$

Example 10.5 Find Δy when x changes from 1 to 2 if:

    a. $dy = x^2\,dx$

    b. $dy = \ln x\,dx$

Solution

    a. $\Delta y = \displaystyle\int_1^2 x^2\,dx = \left.\dfrac{x^3}{3}\right|_1^2 = \dfrac{8}{3} - \dfrac{1}{3} = \dfrac{7}{3}$

    b. $\Delta y = \displaystyle\int_1^2 \ln x\,dx = \left| x\ln x - x \vphantom{\int} \right._1^2 = 2\ln 2 - 1\ln 1$

$$= 2\ln 2$$
$$= 1.39$$

Example 10.6 Evaluate:

    a. $\displaystyle\int_0^2 3e^x\,dx$

    b. $\displaystyle\int_{-1}^2 x^4\,dx$

Solution

    a. $\displaystyle\int_0^2 3\,e^x\,dx = 3\int_0^2 e^x\,dx = \left. 3\,e^x \right|_0^2 = 3\,(e^2 - 1)$

From a table of exponentials, we find that $e^2 = 7.39$. Therefore:

$$\int_0^2 3e^x\,dx = 3\,(7.39 - 1) = 19.2$$

    b. $\displaystyle\int_{-1}^2 x^4\,dx = \left.\dfrac{x^5}{5}\right|_{-1}^2 = \dfrac{2^5}{5} - \dfrac{(-1)^5}{5} = \dfrac{32}{5} + \dfrac{1}{5} = \dfrac{33}{5}$

EXERCISES

1. Evaluate each of the integrals in Exercise number 1 at the end of Section 10.2 between the lower and upper limits respectively:

    a. $x = 1, \quad x = 2$

    b. $x = 3, \quad x = -2$

2. Show algebraically that:

a. $$\int_a^b F(x)\,dx = -\int_b^a F(x)\,dx$$

In other words, show that reversing the limits of a definite integral changes its sign.

b. $$\int_a^c F(x)\,dx = \int_a^b F(x)\,dx + \int_b^c F(x)\,dx$$

10.5 GEOMETRIC INTERPRETATION OF THE DEFINITE INTEGRAL

We saw in Chapter 9 that the derivative, dy/dx, could be interpreted as the slope of the tangent to a plot of $y = f(x)$. The definite integral, like the derivative, can be given a geometric interpretation. Consider the shaded area under the curve of Figure 10.2. It can be shown that this area is equal to the definite integral:

$$\int_a^b F(x)\,dx$$

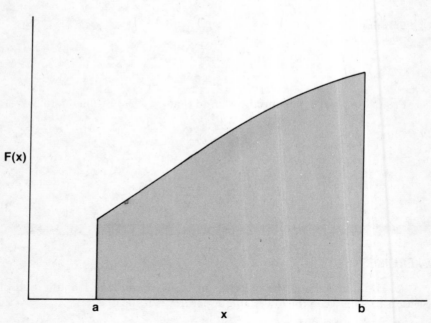

FIGURE 10.2. Area under curve $= \int_a^b F(x)\,dx.$

That is, **the area bounded by the curve of $F(x)$ vs x, the x axis, and the vertical lines $x = a, x = b$ is given by the definite integral of $F(x)\,dx$ between the limits a and b.**

We shall not attempt a rigorous proof of this statement. However, it will be instructive to demonstrate its plausibility by a simple argument that helps us to understand what a definite integral really is. We could estimate the area under the curve in Figure 10.2 by approximating it with the four rectangles shown in Figure 10.3A. Each of these rectangles has a width of Δx. Their heights may be represented as $F(x)_1$, $F(x)_2$, $F(x)_3$, and $F(x)_4$, the ordinates at successive values of x. Clearly:

$$\text{Area of rectangles} = F(x)_1 \Delta x + F(x)_2 \Delta x + F(x)_3 \Delta x + F(x)_4 \Delta x$$

$$= [F(x)_1 + F(x)_2 + F(x)_3 + F(x)_4]\,\Delta x$$

Representing this sum in series notation, we have:

$$\text{Area of rectangles} = \sum_{i=1}^{4} F(x)_i \Delta x$$

Now, the true area under the curve clearly exceeds that of the four rectangles. However, if we were to double the number of rectangles, by making Δx half as large, we would get a better approximation to the true area (Figure 10.3B). That is, the area under the curve is more nearly:

$$\sum_{i=1}^{8} F(x)_i \Delta x', \text{ where } \Delta x' = \Delta x/2$$

It seems reasonable to suppose that if we make Δx smaller and smaller, so as to get a very large number of rectangles, the area of the rectangles would approach more and more closely the area under the curve. In other words:

$$\text{Area under curve} = \operatorname*{Limit}_{\substack{\Delta x \to 0 \\ n \to \infty}} \sum_{i=1}^{n} F(x)_i \Delta x \qquad (10.7)$$

The quantity on the right of Equation 10.7 is precisely what is meant by the definite integral. That is:

$$\int_a^b F(x)\,dx = \operatorname*{Limit}_{\substack{\Delta x \to 0 \\ n \to \infty}} \sum_{i=1}^{n} F(x)_i \Delta x \qquad (10.8)$$

Equation 10.8, which is sometimes referred to as the fundamental theorem of calculus, tells us that the definite integral can be thought of as

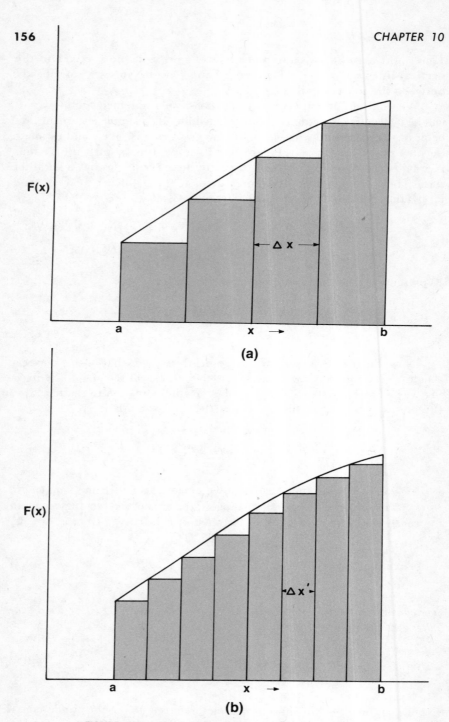

FIGURE 10.3. Successive approximations to area under curve.

the limit of a sum of products:

$$F(x)_1 \Delta x + F(x)_2 \Delta x + \cdots + F(x)_n \Delta x$$

where $F(x)_1$, $F(x)_2 \ldots F(x)_n$ represent ordinates in the region between $x = a$ and $x = b$. From Equations 10.8 and 10.7, we see that:

$$\text{Area under curve} \underset{x=a \to x=b}{} = \int_a^b F(x)\,dx \qquad (10.9)$$

which is the relation we set out to demonstrate.

Example 10.7 Calculate the area between each of the following curves and the x axis.

a. $y = x^2$, from $x = 0$ to $x = 5$
b. $y = \ln x$, from $x = 1$ to $x = 2$
c. $y = x - 2$, from $x = 0$ to $x = 4$

Solution

a. Area $= \displaystyle\int_0^5 x^2\,dx = \left. \dfrac{x^3}{3} \right|_0^5 = \dfrac{125}{3}$

b. Area $= \displaystyle\int_1^2 \ln x\,dx = \left| x \ln x - x \right|_1^2$

$$= (2 \ln 2 - 2) - (0 - 1)$$

$$= 2 \ln 2 - 1$$

$$= 0.386$$

c. Area $= \displaystyle\int_0^4 (x - 2)\,dx = \left| \dfrac{x^2}{2} - 2x \right|_0^4 = 0 - 0 = 0$

This rather puzzling result can be explained if we plot $(x - 2)$ vs x from $x = 0$ to $x = 4$. From Figure 10.4, we see that we are dealing with two areas equal in magnitude but opposite in sign (area I is below the x axis).

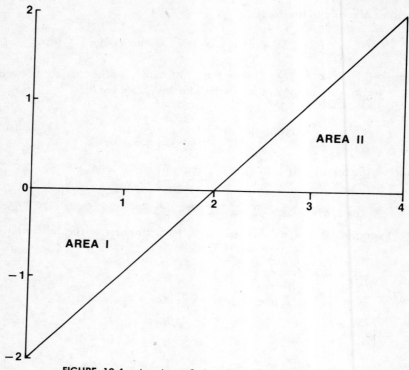

FIGURE 10.4. Area I = −2; Area II = +2; total area = 0.

EXERCISES

1. Evaluate the area between the x axis and the following curves:
   a. $y = 3x - 6$, from $x = 4$ to $x = 6$
   b. $y = 3x - 6$, from $x = 0$ to $x = 2$
   c. $y = e^x$, from $x = 0$ to $x = 1$
2. What is the area between the curve $y = 2x$ and the y *axis* between $y = 0$ and $y = 3$?

PROBLEMS

10.1 A first order reaction is defined as one in which the rate is directly proportional to the concentration of reactant. That is:

$$\frac{-dX}{dt} = kX$$

where X = concentration of reactant, t = time, k = rate constant.

For a first order reaction:

a. Obtain a general relation between X and t. To evaluate the constant of integration, take X_0 to be the value of X when $t = 0$.

b. For a certain first order reaction, $X_0 = 1.00$ and $k = 0.100$ min^{-1}. Calculate X when $t = 10.0$ min.

c. Using the relation derived in part (a), show that the time required for half of the reactant to be consumed is inversely proportional to k and independent of X_0.

10.2 For a second order reaction:

$$\frac{-dX}{dt} = kX^2 \qquad (X = \text{concentration of reactant}, t = \text{time}, k = \text{rate constant})$$

a. Obtain a general relationship between X and t by integrating between the concentration limits X_0 and X and the corresponding time limits 0 and t.

b. If the concentration changes from 1.00 to 0.500 in 10.0 minutes, evaluate the rate constant k.

c. Show that, for a second order reaction, the half life (time required for half of a sample to react) is inversely proportional to both k and X_0.

10.3 A zero order reaction is one in which the rate is a constant, independent of concentration of reactant.

a. Write the differential equation relating concentration (X) and time (t).

b. Obtain the relationship between X and t (take $X = X_0$ when $t = 0$).

c. Show that the half life of a zero order reaction is directly proportional to X_0.

10.4 For the reversible reaction:

$$A \rightleftharpoons B$$

$$\frac{-dX}{dt} = k_1 X - k_2 (X_0 - X)$$

where $X_0 = $ original concentration of A, $X = $ concentration of A at time t, k_1 and k_2 are rate constants for the forward and reverse reactions respectively.

a. Integrate this equation from $X = X_0$, $t = 0$ to X, t, to obtain the relationship between X and t.

    b. Take $X_0 = 1.00$, $k_1 = 0.100$ min^{-1}, $k_2 = 0.200$ min^{-1}. Calculate X when $t = 1, 2, 3,$ and 4 minutes.

    c. For the conditions specified in part (b), what value does X approach as $t \rightarrow \infty$?

10.5 For a certain surface reaction:

$$\frac{-dX}{dt} = \frac{kX}{a - X} \qquad (X = \text{concentration}, t = \text{time}, k = \text{rate constant})$$

Obtain a general expression relating X to t (take $X = X_0$ when $t = 0$).

10.6 The rate of change of vapor pressure (P) with temperature (T) is given by the equation:

$$\frac{dP}{dT} = \frac{\Delta HP}{RT^2} \qquad \begin{array}{l} \Delta H = \text{molar heat of vaporization} \\ R = \text{gas law constant} \end{array}$$

    a. Obtain the general relationship between P and T, involving a constant of integration.

    b. Integrate between the limits (P_1, T_1) and (P_2, T_2) to obtain an equation, without a constant of integration, relating final and initial temperatures and pressures.

10.7 The differential equation relating the fraction of molecules, f, having an energy greater than or equal to E_a, to the absolute temperature, T, is:

$$\frac{df}{dT} = \frac{E_a}{RT^2} e^{-E_a/RT}$$

(R = gas constant, e = base of natural logarithms)

    a. Integrate this equation to obtain the relationship between f and T (let $u = E_a/RT$). To evaluate the constant of integration, note that $f \rightarrow 1$ as $T \rightarrow \infty$.

    b. Obtain an expression for $\dfrac{d \ln f}{dT}$.

10.8 The amount of heat, Q, which must be added to a mole of a gas to increase its temperature from T_1 to T_2 at constant pressure is:

$$Q = \int_{T_1}^{T_2} C_p \, dT$$

where C_p is the constant pressure molar heat capacity. For nitrogen:

$$C_p = 6.52 + 1.25 \times 10^{-3}T - 1.0 \times 10^{-9}T^2 \qquad \text{(in calories)}$$

Calculate Q when T increases from $300°K$ to $500°K$.

10.9   The entropy, S, of a gas changes with temperature according to the relation:

$$\Delta S = \int_{T_1}^{T_2} \frac{C_p}{T}\, dT$$

Using the data in Problem 10.8, calculate ΔS when one mole of nitrogen is heated from $300°K$ to $500°K$.

10.10   The work which is done when a gas expands is given by the relation:

$$W = \int P\, dV \quad (P = \text{pressure}, V = \text{volume})$$

Calculate W when a gas expands from 1.0 liter to 10 liters if:

    a. $P = 1$ atm

    b. $P = \dfrac{24.6}{V}$ lit atm

10.11   Evaluate the integral, $W = \int P\, dV$, given the following experimental data:

V	P	V	P
1.0 lit	20.0 atm	6.0 lit	5.0 atm
2.0	15.0	7.0	4.5
3.0	11.0	8.0	4.3
4.0	8.0	9.0	4.2
5.0	6.0	10.0	4.1

(Plot P vs V and take the area under the curve from $V = 1.0$ lit to 10.0 lit.)

CHAPTER 11

ERROR ANALYSIS

We saw in Chapter 4 that the errors associated with experimental measurements or with quantities calculated from experimental data can be expressed quite simply in terms of significant figures. The rules governing the use of significant figures amount to an elementary form of error analysis, which is sufficient for most of our purposes in general chemistry. Ordinarily, the quantities that we measure in the general chemistry laboratory are based upon a single trial or, at most, upon duplicate determinations. Such experiments hardly justify a more sophisticated type of error analysis than that described in Chapter 4.

We may, however, have the opportunity to carry out more exact experiments where it is important that we be able to estimate quite accurately the uncertainties associated with individual measurements or with results derived from several successive measurements. Let us suppose, for example, that we are asked to determine the percentage of chlorine in a sample of sodium chloride, given the equipment necessary to determine the percentage to four significant figures and the time to make five successive determinations. We might obtain the data listed in Table 11.1.

There are several questions we might raise concerning this data.

1. Since we were able, in principle, to obtain the percentage of chlorine to ±0.01 per cent, why didn't we get the same value in each trial?

2. Should we reject any of these results? (Trial 2 looks a little dubious.)

3. What should we report for the percentage of chlorine? The average of all five trials? The average omitting trial 2? Some other answer?

4. Should we repeat the experiment a few more times to get a "better" value for the percentage of chlorine?

Table 11.1	Percentage of Cl in NaCl				
Trial	1	2	3	4	5
Percentage of Cl	60.50	60.41	60.53	60.54	60.52

5. How confident can we be of the value that we report for the percentage of chlorine? What are the chances that it will be within ±0.01 of the true value? ±0.05?

These are typical of the kinds of questions that we will attempt to find answers for in this chapter. Before proceeding further, it will be necessary to define certain terms that are used repeatedly in error analysis.

11.1   ACCURACY AND PRECISION

The **accuracy** of a measured quantity indicates the extent to which it agrees with the true value. It is described in terms of the **error:**

$$\text{Error} = \text{Observed value} - \text{True value} \qquad (11.1)$$

The smaller the error, the more accurate a measurement is.

The **precision** of a measurement allows us to estimate its reproducibility. It is described by the **deviation,** which is the difference between the observed value and the average value, obtained from a series of measurements:

$$\text{Deviation} = \text{Observed value} - \text{Average value} \qquad (11.2)$$

The smaller the deviation, the more precise the measurement is.

To illustrate the distinction between precision and accuracy, consider the data given in Table 11.1 for the percentage of chlorine in sodium chloride. The average value, which is often referred to as the **arithmetic mean,** is readily found to be 60.50 per cent.

$$\text{Arithmetic mean} = \frac{60.50 + 60.41 + 60.53 + 60.54 + 60.52}{5}$$

$$= \frac{302.50}{5} = 60.50$$

The true value for the percentage of chlorine in sodium chloride is, to four significant figures, 60.66 per cent. (This number is based upon the atomic weights of sodium and chlorine, which have been very carefully determined to *five* significant figures.) Consequently, for the data of Table 11.1, Equations 11.1 and 11.2 become:

$$\text{Error} = \text{Observed value} - 60.66$$

$$\text{Deviation} = \text{Observed value} - 60.50$$

These equations can be used to express the precision and accuracy of each measurement of the percentage of chlorine (Table 11.2).

Table 11.2 Precision and Accuracy of Measured Percentages of Chlorine in NaCl

Trial	Observed value	Error	Deviation	Percent error	Percent deviation
1	60.50	−0.16	0.00	−0.26	0.00
2	60.41	−0.25	−0.09	−0.41	−0.15
3	60.53	−0.13	+0.03	−0.21	+0.05
4	60.54	−0.12	+0.04	−0.20	+0.07
5	60.52	−0.14	+0.02	−0.23	+0.03

Clearly, the precision of these measurements is better than their accuracy. In every trial, the error is greater in magnitude than the deviation. In general, we can always expect the precision of measurements to surpass their accuracy. Only if the arithmetic mean coincides exactly with the true value will our measurements be as accurate as they are precise.

Frequently, we refer to the percent error, or the percent deviation. These quantities are defined as follows:

$$\text{Percent error} = \frac{\text{Error}}{\text{True value}} \times 100 \tag{11.3}$$

$$\text{Percent deviation} = \frac{\text{Deviation}}{\text{Arithmetic mean}} \times 100 \tag{11.4}$$

The numbers in the last two columns at the right of Table 11.2 were calculated in this manner.

Example 11.1 Four different students measure the boiling point of a certain organic liquid at 760 mm Hg pressure. Their results are:

54.9°C, 54.4°C, 54.1°C, and 54.2°C

The true boiling point, to three significant figures, is 54.0°C. Determine, for each measurement, the error, deviation, percent error, and percent deviation.

Solution The error can be calculated directly from Equation 11.1. In order to use Equation 11.2 to calculate the deviation, we must first obtain the arithmetic mean. We could do this by adding the four numbers and dividing the sum by four. To reduce the amount of arithmetic, we might write:

$$\text{Arithmetic mean} = 54.0 + \frac{0.9 + 0.4 + 0.1 + 0.2}{4}$$

$$= 54.0 + \frac{1.6}{4} = 54.4$$

For the first measurement, we have:

$$\text{Error} = 54.9 - 54.0 = 0.9;$$

$$\text{Percent error} = \frac{0.9}{54.0} \times 100 = +2\%$$

$$\text{Deviation} = 54.9 - 54.4 = 0.5;$$

$$\text{Percent deviation} = \frac{0.5}{54.4} \times 100 = +0.9\%$$

Proceeding similarly for each measurement:

Observed value	Error	Percent error	Deviation	Percent deviation
54.9	0.9	+2	0.5	+0.9
54.4	0.4	+0.7	0.0	0.0
54.1	0.1	+0.2	−0.3	−0.6
54.2	0.2	+0.4	−0.2	−0.4

EXERCISES

1. Repeat the calculations in Example 11.1 for temperatures expressed in °K (°K = °C + 273.2°).

2. In 1894, Lord Rayleigh determined the mass of nitrogen gas filling a certain container at a known pressure and temperature. The results of successive weighings were as follows:

2.3102 g,   2.3099 g,   2.3101 g,   2.3100 g,   2.3102 g,   2.3101 g,
2.3103 g,   2.3116 g,   and   2.3096 g

The nitrogen he used was prepared from air by removing oxygen, water vapor, and carbon dioxide. When he performed similar experiments with chemically pure nitrogen, he obtained a mass of 2.2997 g. Taking this to be the "true" mass, calculate the deviations and errors of each of the masses listed above.

11.2   TYPES OF ERRORS

The most serious errors that students make in the chemistry laboratory (or in life, for that matter) are ones which could either be avoided or corrected for. These are called **determinate errors.** As an example, consider a student who is attempting to analyze a metal oxide by heating it in a stream of hydrogen to form the pure metal.

$$MO(s) + H_2(g) \rightarrow M(s) + H_2O(g)$$

If he spills part of his sample, his result is likely to show a rather large error. This error could be avoided by being more careful; the only way to "correct" for it would be to repeat the experiment.

As an example of a determinate error which could be corrected for quite readily, consider the dilemma of a student who determines the density of benzene at 20°C by weighing samples of the liquid issuing from a 10 ml pipette. He obtains the following masses:

8.681 g, 8.678 g, 8.683 g, and 8.678 g

Finding the arithmetic mean of these masses to be 8.680 g, he divides by 10.00 ml and confidently reports a density of 0.8680 g/ml, trusting that his excellent precision insures high accuracy. Unfortunately, the true value is 0.8790 g/ml; the student has made an error of more than 1 per cent. He failed to realize that a 10 ml pipette does not deliver exactly ten ml of liquid. Had he taken the trouble to calibrate his pipette with distilled water, he would have found that it delivered about 9.90 ml. He would then have reported:

density benzene = 8.680 g/9.90 ml = 0.877 g/ml

with an error of about 0.2 per cent.

Example 11.2   A student determines the gram equivalent weight of a metal by reducing it with hydrogen. He makes three weighings:

test tube	mass = A
test tube + metal oxide	mass = B
test tube + metal	mass = C

and calculates the gram equivalent weight using the equation:

$$\text{G.E.W.} = 8.000 \text{ g O} \times \frac{\text{mass metal}}{\text{mass oxygen}} = 8.000 \text{ g O} \times \frac{(C - A)}{(B - C)}$$

How will each of the following determinate errors affect the accuracy of his result (i.e., will they make it larger or smaller than the true value)?

    a. In weighing the test tube, he records a mass one gram greater than the true mass.

    b. After weighing the metal oxide plus test tube, he spills part of the oxide.

    c. He fails to convert all of the metal oxide to metal.

Solution

    a. Mass A is too large; B and C are presumably correct. From the equation for G.E.W., we see that the calculated mass of metal, $C - A$, will be too small. Hence, the G.E.W., which is directly proportional to $(C - A)$, will be too small.

    b. Here, A and B will be correct. Mass C will be too small, because sample is lost between the second and third weighings. Looking at the equation for G.E.W., we see that $(C - A)$ will be too small, $(B - C)$ will be too large, and the calculated G.E.W. will be too small.

    c. Again, A and B are correct; C is too large because not all of the sample is reduced. Following the reasoning of part (b), we deduce that the calculated G.E.W. will be too large.

Determinate errors account for the fact that the accuracy of experiments rarely equals their precision. We find, however, that even when all determinate errors are eliminated, a series of measurements shows a certain amount of scatter, reflected in deviations from the mean. These deviations are due to what are called **indeterminate errors.** The adjective "indeterminate" implies that these errors cannot be corrected for. In other words, we cannot in any rational way adjust our results to compensate for or to eliminate errors of this type.

Indeterminate errors result from inherent imperfections in the instruments and techniques used to carry out measurements. To illustrate this point, consider the student who determines the density of benzene using a carefully calibrated pipette and a high quality analytical balance. In four trials, he might obtain densities of:

    0.8782 g/ml,   0.8794 g/ml,   0.8785 g/ml, and   0.8779 g/ml

The indeterminate errors that are responsible for deviations in this series of experiments could be due to:

(a) Slight fluctuations in temperature; the density of a liquid varies with temperature.

(b) Evaporation of a small amount of benzene before it is weighed.

(c) Failure to make the benzene level in the pipette coincide exactly with the graduation mark.

(d) ? ? ?

Even though indeterminate errors cannot be corrected for, they can be treated statistically to tell us how they are likely to affect the reliability of our measurements. This type of analysis is based on the so-called normal error curve shown in Figure 11.1. This curve tells us the relative frequency of the various deviations that we can expect to find if we make a large number of measurements. The figure allows us to make some general observations about the magnitude of indeterminate errors.

1. Since the curves are symmetric about the midpoint, representing the arithmetic mean, positive and negative deviations are equally likely.

2. Since the curves rise to a maximum at the midpoint, small deviations occur more frequently than large deviations.

3. The shape of the curve is determined by the inherent precision of the measurement. If the instruments or techniques that we use are

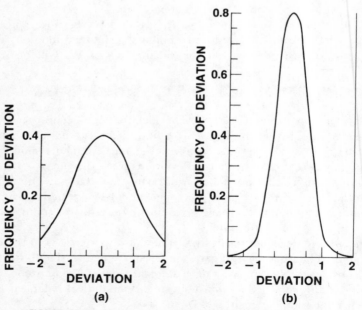

FIGURE 11.1. The error curve: (a), low precision; (b), high precision.

relatively crude or "sloppy," we can expect to have an error curve of the type shown at the left of Figure 11.1, with a relatively high frequency of large deviations. As we refine our measurements to improve their precision, we would expect to approach the error distribution shown at the right of Figure 11.1, where large deviations are highly improbable.

EXERCISES

1. A student determining the G.E.W. of a metal makes the following weighings:

$$\text{crucible} \qquad\qquad \text{mass} = A$$
$$\text{crucible} + \text{metal} \qquad\qquad \text{mass} = B$$
$$\text{crucible} + \text{metal sulfide} \qquad\qquad \text{mass} = C$$

$$\text{G.E.W.} = 16.0 \text{ g S} \times \frac{\text{mass metal}}{\text{mass sulfur}} = 16.0 \text{ g S} \times \frac{(B - A)}{(C - B)}$$

How will each of the following determinate errors affect the accuracy of his calculated G.E.W.?

    a. After weighing the crucible plus metal, he spills some of the metal.

    b. Some of the metal sulfide (MS) is oxidized to sulfate (MSO_4).

    c. The mass of the crucible plus metal sulfide is incorrectly recorded as 16.900 g; it should have been 16.100 g.

2. Using Figure 11.1, estimate the relative frequency of obtaining deviations of 0.5 vs 1.0 for measurements associated with inherently

    a. low precision (Figure 11.1a).

    b. high precision (Figure 11.1b).

3. A class of 20 students measures the length of a certain spectral line with the following results:

2050 Å	2050 Å	2054 Å	2052 Å	2050 Å
2049	2051	2047	2049	2049
2051	2048	2050	2048	2051
2053	2052	2051	2045	2050

    a. Find the arithmetic mean and the deviation of each measurement.

    b. Construct an "error curve" by plotting the frequency (number) of each deviation vs the magnitude of the deviation.

11.3 MEASURES OF PRECISION

Two quite different quantities are used to describe the scatter of experimental measurements: the **average deviation** and the **standard deviation.** The average deviation is readily calculated by taking the sum of the deviations, regardless of sign, from the arithmetic mean, and dividing it by the total number of observations.

$$a = \frac{\Sigma \, |d|}{n} \tag{11.5}$$

where a is the average deviation, $|d|$ represents the magnitude of an individual deviation, neglecting its sign, and n is the number of trials.

Example 11.3 Five students report the following percentages of chlorine in a sample:

$$19.82, \ 19.57, \ 19.68, \ 19.71, \text{ and } 19.75$$

Calculate the arithmetic mean of these results and the average deviation.

Solution To obtain the arithmetic mean, it is convenient to take 19.50 as our base number and write:

$$\text{Arithmetic mean} = 19.50 + \frac{0.32 + 0.07 + 0.18 + 0.21 + 0.25}{5}$$

$$= 19.50 + \frac{1.03}{5} = 19.71$$

(Some people prefer to carry one extra digit in the mean; in this case, they would write 19.706 for the mean. We shall round off to 19.71 to simplify the arithmetic.) We proceed to calculate the individual deviations, d, and the magnitude of each deviation, $|d|$.

Trial	1	2	3	4	5
Percentage of Cl	19.82	19.57	19.68	19.71	19.75
d	+0.11	−0.14	−0.03	0.00	+0.04
$|d|$	0.11	0.14	0.03	0.00	0.04

$$a = \frac{0.11 + 0.14 + 0.03 + 0.00 + 0.04}{5} = 0.06_4$$

The average deviation gives us a qualitative estimate of the precision of our data. Unfortunately, it has no direct statistical significance. The **standard deviation** is a more significant quantity in that it determines the shape of the error curve to be associated with a series of measurements. Unlike the average deviation, it cannot be calculated exactly from a limited amount of experimental data. It can, however, be estimated from the approximate relation:

$$\sigma = \sqrt{\frac{\Sigma d^2}{n - 1}} \tag{11.6}$$

where σ is the standard deviation and, as before, d is an individual deviation and n is the number of trials.

Example 11.4 Estimate the standard deviation for the data given in Example 11.3.

Solution We recall that the arithmetic mean is 19.71. Calculating deviations from the mean, we have:

Trial	1	2	3	4	5
Percentage of Cl	19.82	19.57	19.68	19.71	19.75
d	+0.11	−0.14	−0.03	+0.00	+0.04
d^2	0.0121	0.0196	0.0009	0.0000	0.0016

$$\Sigma d^2 = 0.0121 + 0.0196 + 0.0009 + 0.0000 + 0.0016 = 0.0342$$

$$n - 1 = 5 - 1 = 4$$

$$\Sigma d^2 / (n - 1) = 0.0342/4 = 0.0086$$

$$\sigma = (0.0086)^{1/2} = 0.09_2$$

Comparing the answers to Examples 11.3 and 11.4, we see that the standard deviation, 0.092, is greater than the average deviation, 0.066. This is ordinarily the case, which may explain why many people prefer to use the average deviation rather than the standard deviation to report the precision of their results! It can be shown that, for a very large number of trials, the standard deviation approaches $5/4$ of the average deviation:

$$\sigma \to \frac{5}{4}\, a, \text{ as n} \to \infty \qquad \text{(n = number of trials)} \tag{11.7}$$

The importance that we attach to the standard deviation reflects the influence it has on the shape of the error curve. It can be shown (see Exercise 4 at the end of this section) that a small value of σ corresponds to a sharp, steeply rising curve, on which the vast majority of deviations are very close to zero. Conversely, a large σ leads to a broad, squat error curve, on which large deviations have a relatively high probability.

In Figure 11.2, we have shown a general form of the error curve similar to those shown in Figure 11.1, except that the divisions along the horizontal axis are expressed as multiples of the standard deviation, σ ($x = -3\sigma, -2\sigma, -\sigma, 0, \sigma, 2\sigma, 3\sigma$). To interpret this curve, let us look at the shaded area, which is bounded on the left and right respectively by the vertical lines $x = -\sigma$ and $x = \sigma$. *This area is proportional to the probability of observing a deviation within one unit of σ of the arithmetic mean*, located at the midpoint of the curve. The shaded area comprises a little more than $\frac{2}{3}$ of the total area under the error curve. This means that if we were to make a large number of trials, we would expect about $\frac{2}{3}$ of them to fall within the range:

$$M - \sigma \text{ to } M + \sigma, \text{ or } M \pm \sigma$$

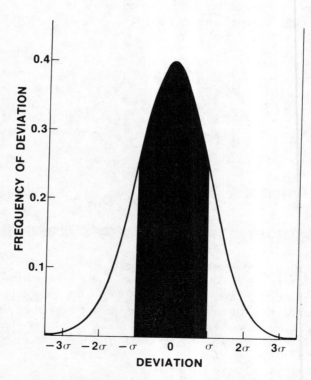

FIGURE 11.2. Normal error curve.

where M is the arithmetic mean and σ is the standard deviation. Slightly less than $\frac{1}{3}$ of the trials would show larger deviations and hence would fall outside of this range.

As we move farther away from the midpoint of the error curve, let us say to $\pm 2\sigma$, we enclose a still greater portion of the total area. The region between $x = -2\sigma$ and $x = +2\sigma$ comprises about 95 per cent of the total area. We interpret this to mean that if we made a large number of measurements, we would expect about 95 per cent of them to fall in the range:

$$M - 2\sigma \quad \text{to} \quad M + 2\sigma, \quad \text{or} \quad M \pm 2\sigma$$

Only about 5 per cent, or 1 in 20, would fall outside this range; i.e., would show a deviation greater in magnitude than 2σ.

Table 11.3 Probability of a Result Falling within a Specified Range of The Arithmetic Mean (M)

Range	Within the range (percentage)	Outside the range (percentage)
$M \pm 0.674\ \sigma$	50.00	50.00'
$M \pm 1.000\ \sigma$	68.3	31.7
$M \pm 1.645\ \sigma$	90.00	10.00
$M \pm 2.000\ \sigma$	95.5	4.5
$M \pm 2.576\ \sigma$	99.00	1.00
$M \pm 3.000\ \sigma$	99.7	0.3
$M \pm 3.291\ \sigma$	99.90	0.1

One feature of Table 11.3 is particularly noteworthy. It can be seen that half of the deviations can be expected to fall in the range:

$$M - 0.674\sigma \quad \text{to} \quad M + 0.674\sigma \quad (M \pm 0.674\sigma)$$

The quantity 0.674σ is sometimes referred to as the probable deviation:

$$p = 0.674\sigma \tag{11.8}$$

with the understanding that there is a 50-50 chance that a deviation, chosen at random, would exceed the probable deviation, p.

Example 11.5 Referring to Example 11.4 and Table 11.3:
a. Assuming that a large number of students determine the percentage of chlorine in a sample, within what range would you expect 90 per cent of the results to fall?

b. What percentage of the class would you expect to report results greater than 19.80 per cent?

Solution

a. 90 per cent of the results should fall in the range (Table 11.3):

$$M - 1.645\sigma \quad \text{to} \quad M + 1.645\sigma \quad (M \pm 1.645\sigma)$$

In Example 11.4, we found M = 19.71, σ = 0.092. This range must then be:

$$19.71 \pm 1.645 \, (0.092) = 19.71 \pm 0.15$$

In other words, 90 per cent of the students would be expected to submit results between 19.56 and 19.86.

b. 19.80 falls about one standard deviation *above* the mean. We would expect about 32 per cent of the students to report results deviating from the mean by more than one standard deviation. Of these, half would be on the high side (>19.80) and half on the low side (<19.62). We deduce that about one half of 32 per cent, or 16 per cent of the students would report results higher than 19.80 per cent.

A word of caution is in order concerning the kind of predictions made in Example 11.5. Like any predictions made on the basis of the laws of probability, *they can be expected to be valid only when we are dealing with a large number of observations*. With only a few results to work with, we can expect, at best, an error pattern which crudely approximates that shown in Figure 11.2 or Table 11.3. Indeed, if we have only a few observations, we cannot expect to obtain the true arithmetic mean, M, on which the figure and table are based. It is highly likely that if more results were available, the apparent mean would change to a new value, hopefully approaching more and more closely the true mean of an infinite number of trials.

EXERCISES

1. Calculate a and σ for the data in Table 11.1.
2. A careful experimenter obtains the following values for the atomic weight of cadmium: 112.25, 112.36, 112.32, 112.21, 112.30, and 112.36.
   a. What are the arithmetic mean and the standard deviation?

  b. If the experimenter makes one more measurement, what is the likelihood that it will fall within ±0.06 of the mean?
  c. If many more trials are made, within what range would you expect 50 per cent of them to fall?
3. Referring to Example 11.5:
  a. Within what range would you expect 99 per cent of the results to fall?
  b. What percentage of a class would you expect to report results less than 19.53 per cent?
4. The equation of the error curve is:

$$y = \frac{e^{-x^2/2\sigma^2}}{2.5\sigma}$$

Plot a curve of y vs x in the region $x = -2$ to $x = +2$ when:
a. $\sigma = 1$
b. $\sigma = 0.5$

11.4 RELIABILITY OF THE MEAN

When we carry out a quantitative experiment, we are always restricted to a limited number of trials, often as few as two, seldom more than five or six. The value which we report is ordinarily the arithmetic mean of the several trials. The question arises as to how much confidence we can place in this value. If we could make an infinite number of determinations, how close could we reasonably expect the true mean to come to the mean calculated on the basis of a few trials?

Questions such as this can be answered statistically in terms of what are known as **confidence levels.** Consider, for example, Exercise 2 at the end of Section 11.3, in which we showed that the mean value of the atomic weight of cadmium, based on six trials, was 112.30. It can be shown statistically (see Exercise 1 at the end of this section) that the true mean will fall within ±0.02 of 112.30 at the "50 per cent confidence level," which we interpret to mean that there is a 50-50 chance that the true mean would fall in the range 112.28–112.32. Again, we can show that the true mean will fall within ±0.05 of 112.30 at the "90 per cent confidence level." In other words, the chances are 9 in 10 that the true mean, which could be determined only by making an infinite number of trials, would fall between 112.25 and 112.35.

Reliability limits can be assigned to a calculated mean by using the so-called "t-test," which makes use of the equation:

$$M = \text{Calculated mean} \pm t\sigma/\sqrt{n} \qquad (11.9)$$

where M is the true mean, σ is the standard deviation, n is the number of trials, and t is a statistical parameter which can be obtained from tables such as Table 11.4.

	Table 11.4	**Values of t in Equation 11.9**	
		Confidence level	
n	**50 per cent**	**90 per cent**	**99 per cent**
2	1.000	6.314	63.657
3	0.816	2.920	9.925
4	0.765	2.353	5.841
5	0.741	2.132	4.604
6	0.727	2.015	4.032
8	0.711	1.895	3.500
10	0.703	1.833	3.250
∞	0.674	1.645	2.576

To illustrate how Equation 11.9 is used, let us consider a specific example. Suppose a student, asked to determine the molecular weight of an organic solute, makes three trials with the following results:

$$159.0, \ 161.0, \text{ and } 160.0$$

Clearly, the calculated mean is 160.0. The standard deviation (Equation 11.6) is:

$$\sigma = \sqrt{\frac{\Sigma d^2}{n-1}} = \sqrt{\frac{(1.0)^2 + (1.0)^2}{2}} = 1.0$$

Substituting in Equation 11.9, we obtain:

$$M = 160.0 + t(1.0)/\sqrt{3} = 160.0 + 1.0t/1.7 = 160.0 + 0.59t$$

Reading across Table 11.4 at $n = 3$:

50 per cent level: $M = 160.0 \pm 0.816(0.59)$

$\qquad\qquad\qquad = 160.0 \pm 0.5;\qquad$ range $= 159.5 - 160.5$

90 per cent level: $M = 160.0 \pm 2.920(0.59)$

$\qquad\qquad\qquad = 160.0 \pm 1.7;\qquad$ range $= 158.3 - 161.7$

99 per cent level: $M = 160.0 \pm 9.925(0.59)$

$\qquad\qquad\qquad = 160.0 \pm 5.8;\qquad$ range $= 154.2 - 165.8$

This calculation tells us that if an infinite number of trials were made, there is a 50-50 chance that the true mean would fall within ± 0.5 of 160.0, 9 chances in 10 that it would fall within ± 1.7 of 160.0, and 99 chances in 100 that it would be within ± 5.8. Putting it another way, there is one chance in two that the true mean will fall *outside* the range 159.5–160.5; 1 chance in 10 that it will fall outside the range 158.3–161.7, and only one chance in 100 that it will be less than 154.2 or greater than 165.8.

As this example illustrates, the higher the confidence level we demand, the less precisely we can specify the value of the true mean. To choose two extreme cases: we can be 0 per cent confident that the true mean coincides exactly with the calculated mean ($M = 160.0 \pm 0.0$), but we can be 100 per cent confident that the true mean falls between $+ \infty$ and $- \infty$!

Example 11.6 The observed mean of a series of determinations of the density of a liquid is 1.69 g/ml. Calculate reliability limits at the 90 per cent confidence level if:

    a. $\sigma = 0.10, n = 2$
    b. $\sigma = 0.05, n = 2$
    c. $\sigma = 0.05, n = 5$

Solution

    a. From Table 11.4, we find that for $n = 2$ at the 90 per cent confidence level, $t = 6.314$

$$M = 1.69 \pm \frac{6.3\,(0.10)}{\sqrt{2}} = 1.69 \pm 0.45$$

    b.
$$M = 1.69 \pm \frac{6.3\,(0.05)}{\sqrt{2}} = 1.69 \pm 0.22$$

    c. $t = 2.132;$
$$M = 1.69 \pm \frac{2.1\,(0.05)}{\sqrt{5}} = 1.69 \pm 0.05$$

We see from Example 11.6 (and from Equation 11.9) that the reliability of the mean depends upon the magnitude of the standard deviation. Large values of σ imply poor precision and lead to a relatively unreliable mean. The reliability of the calculated mean is also a function of the number of observations upon which it is based. The more trials we make, the more confidence we have that the mean we calculate is a good approximation to the true mean. In Example 11.6, we found that when n increased from 2 to 5, the limits of M, at the 90 per cent confidence level, sharpened from ± 0.22 to ± 0.05.

It should be noted, however, that beyond a certain point, there is little to be gained by increasing the number of observations. Referring again to Example 11.6, let us see what happens if we make $n = 6$ when $\sigma = 0.05$. From Table 11.4, we find $t = 2.015$ at the 90 per cent level. Hence:

$$M = 1.69 \pm \frac{2.0\,(0.05)}{\sqrt{6}} = 1.69 \pm 0.04$$

Comparing this result with that calculated in Example 11.6, part (c), we see that increasing n from 5 to 6 had very little effect on the reliability limits of M (± 0.04 instead of ± 0.05).

Strictly speaking, Equation 11.9 tells us only the deviation which we may reasonably expect between the observed and true mean. *If* we can assume that there are no determinate errors involved, the true mean should coincide with the true value of the quantity we are measuring. In this case, Equation 11.9 should tell us the expected *error* to be associated with the mean. Thus in Example 11.6, part (c), if the precision and accuracy of the density measurements are the same, we could report the density to be 1.69 ± 0.05 with 90 per cent confidence that the true density would lie within the indicated limits.

Equation 11.9 can be used in this manner to calculate what is known as the **probable error** of the mean.

$$\text{P.E.} = \frac{t\sigma}{\sqrt{n}} \qquad \text{(50 per cent confidence level)} \qquad (11.10)$$

In principle, there is a 50-50 chance that the error of the mean will be less than that calculated by Equation 11.10. In practice, the chances are considerably less than even that the true value will fall within the indicated range, primarily because the accuracy of the measurement is almost certain to be poorer than the precision.

The expression for the probable error is often further simplified by taking the t value to be that at $n = \infty$ (Table 11.4) and rewriting Equation 11.10 as:

$$\text{P.E.} \approx \frac{0.67\sigma}{\sqrt{n}} \qquad (11.11)$$

Equation 11.11 gives us an estimate of the probable error of the mean which is even more optimistic than that in Equation 11.10, since t for a finite number of observations will be greater than 0.67.

EXERCISES

1. Consider the data of Exercise 2, Section 11.3. What are the reliability limits of the mean at the 50 per cent, 90 per cent, and 99 per cent confidence limits?

2. A student, in two trials, finds the values 16.6 and 17.2 for the percentage of chlorine in a sample. If he wishes to be 99 per cent sure of his answer, what limits should he put on the value he reports?

3. A student finds the following values for the gram equivalent weight of an acid:

$$202.4, 199.8, 201.7, \text{ and } 200.8$$

Calculate the reliability limits of the mean at the 50 per cent confidence level and compare to the probable error calculated from Equation 11.11.

11.5 REJECTION OF A RESULT

Let us suppose that a student in the general chemistry laboratory, asked to find the molecular weight of an organic liquid, makes five determinations with the following results:

$$154.2, 148.0, 153.5, 152.9, \text{ and } 154.5$$

So far as he knows, each run was carried out in the same manner. There was no obvious determinate error in any of the trials. Yet, the number 148.0 appears to be out of line; should it be rejected in calculating the mean?

The question we are really asking is, "Does the value 148.0 reflect a determinate error of which the student was unaware?" If this is the case, it should be rejected. To answer this question, we must ask another. What are the chances that a deviation of this magnitude would turn up normally in carrying out a series of experiments? Unfortunately, there are no hard and fast rules which will give us definitive answers to these questions. Chemists and other scientists frequently employ one or another of three empirical rules to decide when to reject a doubtful result.

1. The 2.5a rule. The value is rejected if its deviation from the trial mean, calculated by ignoring the doubtful value, is greater than 2.5 times the average deviation, a. Applying this rule to the molecular weight data discussed above, we first take the mean of the four "reliable" values:

$$\frac{154.2 + 153.5 + 152.9 + 154.5}{4} = 153.8$$

$$a = \frac{\Sigma \mid d \mid}{n} = \frac{0.4 + 0.3 + 0.9 + 0.7}{4} = 0.6$$

The deviation of the suspected value, 148.0, from the mean, 153.8 is 5.8. Since

$$5.8 > 2.5\,(0.6) = 1.5$$

this rule would clearly reject the value 148.0.

2. The 4a rule. This rule is similar to the rule given in (1), except that the criterion for rejection is $4a$ rather than $2.5a$. Clearly, if we use this rule instead of (1), we are less likely to reject doubtful values. If the deviation of the suspected value were 3 times the average deviation, it would be retained if we used the $4a$ rule and rejected if we used the $2.5a$ rule.

In the case of the molecular weight data, the $4a$ rule would advise us to reject the number 148.0.

$$5.8 > 4\,(0.6) = 2.4$$

3. The Q test. This test, which is used more frequently today than either of the preceding rules, is based on the following procedure:

(a). Arrange the results in ascending order:

Example: 148.0, 152.9, 153.5, 154.2, 154.5

(b). Obtain the difference between the suspected value and the value nearest to it:

$$152.9 - 148.0 = 4.9$$

(c). Calculate the range, i.e., the difference between the highest and lowest value, including the suspect:

$$154.5 - 148.0 = 6.5$$

(d). Obtain a quotient, Q, by dividing the answer obtained in step (b) by the answer obtained in (c):

$$Q = 4.9/6.5 = 0.75$$

(e). Compare to the value of Q found in Table 11.5, or similar tables. If the calculated Q exceeds that given in the table, reject the suspected value. Looking at Table 11.5, we find that $Q_{0.90}$ for $n = 5$ (5 observations) is 0.64. Since 0.75 is greater than 0.64, we would reject the number 148.0.

n	3	4	5	6	8	10
Table 11.5 The Q Test						
$Q_{0.90}$	0.94	0.76	0.64	0.56	0.47	0.41

This rule tends to be more stringent than either rule (1) or rule (2) in the sense that it is less likely to advise us to reject a suspected value (see Example 11.7).

Each of the rules that we have stated is based upon statistical considerations of one type or another. The Q test, in particular, can be applied with 90 per cent confidence that the rejected value is significantly different from the others. That is, 90 per cent of the values rejected on the basis of the Q test reflect determinate errors rather than normal statistical fluctuations. Each of the rules suffers from one fundamental defect. With a limited number of observations, we are not in a position to accurately estimate the true mean, and hence the true deviation, of the suspected value. In the case of the molecular weight determinations, for example, it is quite possible that if we were to carry out ten determinations rather than five, we might obtain the following results:

$$154.2, \ 148.0, \ 153.5, \ 152.9, \ 154.5$$

$$149.6, \ 152.0, \ 148.4, \ 154.8, \ 151.2$$

in which case the trial mean would shift from 153.8 to 152.3, and one can readily show (see Exercise 1 at the end of this section) that each of the three rules would advise us to *retain* the number 148.0.

Example 11.7 A student obtains the following results for the molarity of an NaOH solution: 0.504, 0.510, 0.514, and 0.538. Apply the $4a$ rule and the Q test to decide whether the number 0.538 should be rejected.

Solution

$4a$ rule: trial mean $= \dfrac{0.504 + 0.510 + 0.514}{3} = 0.509$

$$a = \frac{\Sigma \ |\ d\ |}{n} = \frac{0.005 + 0.001 + 0.005}{3} = 0.004$$

Deviation of suspected value $= 0.029$

$0.029 > 4\,(0.004) = 0.016$ *Reject*

Q test:

$$\begin{array}{cccc} & & \overset{0.024}{\longleftrightarrow} & \\ 0.504 & 0.510 & 0.514 & 0.538 \\ & \underset{0.034}{\longleftrightarrow} & & \end{array}$$

$Q = 0.024/0.034 = 0.71$

From Table 11.5, we find $Q_{0.90} = 0.76$ for $n = 4$.
Since $Q < Q_{0.90}$, we should *retain* the value 0.538.

Not infrequently, we find that, with a limited number of trials, the Q test advises us to retain a value which the $2.5a$ rule or the $4a$ rule would reject. What do we do in such a case? The obvious answer is to carry out more determinations, in which case the three rules become more reliable and more nearly consistent. If we cannot do this, we have to accept the fact that statistical considerations are not very helpful when we are limited to a few trials. We may well find ourselves with the dilemma of choosing between the Q test, which is likely to retain invalid results, and the $4a$ and $2.5a$ rules, which are successively more likely to reject valid results.

EXERCISES

1. Apply the $2.5a$ rule, the $4a$ rule, and the Q test to the ten molecular weight values given on p. 181 to see whether 148.0 should be rejected.

2. A student determines the percentage of iron in an ore, obtaining six values:

$$47.0, \ 55.2, \ 49.4, \ 50.1, \ 49.6, \ \text{and} \ 50.5$$

Apply the Q test successively to the highest and lowest numbers to see if any of them should be rejected. That is, test 55.2 first, then 47.0, then, if necessary, 50.5, and so on.

11.6 ERROR OF A CALCULATED RESULT

Most of the quantities that we determine in the laboratory are based upon calculations involving more than one measured quantity. Associated with each individual measurement, there will be a certain error which may be estimated, perhaps by a formula such as that given by Equation 11.11. How should these errors be combined to estimate the total error to be associated with a calculated result?

In Chapter 4, we described how this question could be answered in an approximate way by using the rules of significant figures. A more exact treatment leads to the rules given in Table 11.6.

Table 11.6   Error in a Derived Result

Operation	Example	Determinate error	Indeterminate error
1. Addition	$S = A + B$	$\lvert \varepsilon_s \rvert = \lvert \varepsilon_a \rvert + \lvert \varepsilon_b \rvert$	$E_s^2 = E_a^2 + E_b^2$
2. Subtraction	$D = A - B$	$\lvert \varepsilon_d \rvert = \lvert \varepsilon_a \rvert + \lvert \varepsilon_b \rvert$	$E_d^2 = E_a^2 + E_b^2$
3. Multiplication	$P = A \times B$	$\left\lvert \dfrac{\varepsilon_p}{P} \right\rvert = \left\lvert \dfrac{\varepsilon_a}{A} \right\rvert + \left\lvert \dfrac{\varepsilon_b}{B} \right\rvert$	$\left(\dfrac{E_p}{P} \right)^2 = \left(\dfrac{E_a}{A} \right)^2 + \left(\dfrac{E_b}{B} \right)^2$
4. Division	$Q = A/B$	$\left\lvert \dfrac{\varepsilon_q}{Q} \right\rvert = \left\lvert \dfrac{\varepsilon_a}{A} \right\rvert + \left\lvert \dfrac{\varepsilon_b}{B} \right\rvert$	$\left(\dfrac{E_q}{Q} \right)^2 = \left(\dfrac{E_a}{A} \right)^2 + \left(\dfrac{E_b}{B} \right)^2$

The equations listed in Column 3 of Table 11.6 give the errors to be expected in a calculated result when the errors in A and B, $\lvert E_a \rvert$, and $\lvert E_b \rvert$ are determinate in origin. (Note that we are concerned with the absolute magnitude of these errors and not with their signs.) In addition and subtraction, the total error is the sum of the errors in the individual quantities. For multiplication and division, the fractional (or percent) error in the sum or product is equal to the sum of the *fractional* (or percent) errors in A and B (Example 11.8).

Example 11.8   The density of a liquid is determined by finding the mass and volume of a sample and applying the defining equation:

$$\rho = m/V$$

If the mass is known to be in error by 0.1 per cent and the volume by 1.0 per cent, what is the percent error in ρ?

Solution   If the percent error in m is 0.1, then: $\dfrac{\varepsilon_m}{m} = 0.001$

Similarly: $\dfrac{\varepsilon_v}{V} = 0.010$

Hence: $\varepsilon_\rho/\rho = 0.001 + 0.010 = 0.011$

Percent error $= 100\,\varepsilon_\rho/\rho = 1.1\%$

When the errors associated with measured quantities are indeterminate in origin, we use Column 4 of Table 11.6 to estimate the error in a derived result.

Example 11.9 A student determines the gram equivalent weight of iron by reducing a weighed sample of an oxide of iron to the metal and using the equation:

$$\text{G.E.W. Fe} = 8.000 \text{ g O} \times \frac{\text{wt. iron}}{\text{wt. oxygen}}$$

His weighings, along with the estimated *indeterminate* errors, are as follows:

empty test tube: $12.602 \pm 0.002 \text{ g}$
test tube + iron oxide: $14.709 \pm 0.002 \text{ g}$
test tube + iron: $14.076 \pm 0.002 \text{ g}$

He calculates the gram equivalent weight to be:

$$8.000 \text{ g} \times \frac{1.474 \text{ g}}{0.633 \text{ g}} = 18.63 \text{ g}$$

Estimate the error in:
a. The weight of iron.
b. The weight of oxygen.
c. The G.E.W. of iron.

Solution

a. Using the last column in Table 11.6:

$$E^2 = (2 \times 10^{-3})^2 + (2 \times 10^{-3})^2$$
$$= 4 \times 10^{-6} + 4 \times 10^{-6} = 8 \times 10^{-6}$$
$$E = (8 \times 10^{-6})^{1/2} = 2.8 \times 10^{-3} \text{ g}$$

b. Here again, two weighings, each with an error of 2×10^{-3} g, are involved:
Hence: $E = 2.8 \times 10^{-3}$ g

c. The gram equivalent weight is calculated by taking the ratio of two weights, each of which is in error by 2×10^{-3} g:

$$\left(\frac{E}{18.63}\right)^2 = \left(\frac{2.8 \times 10^{-3}}{1.474}\right)^2 + \left(\frac{2.8 \times 10^{-3}}{0.633}\right)^2$$
$$= 3.6 \times 10^{-6} + 20 \times 10^{-6} = 24 \times 10^{-6}$$

$$\frac{E}{18.63} = (24 \times 10^{-6})^{1/2} = 4.9 \times 10^{-3}$$

$$E = (0.0049)(18.63) = 0.09$$

We conclude that: G.E.W. $= 18.63 \pm 0.09$

It will be noted from Example 11.9 that the fractional error in the gram equivalent weight (E/18.63) can be attributed almost entirely to that in the weight of oxygen ($2.8 \times 10^{-3}/0.633$); the smaller fractional error in the weight of iron ($2.8 \times 10^{-3}/1.474$) made a relatively small contribution to the total error. It is generally true that where indeterminate errors are involved, it is the error in the least accurate quantity that determines the magnitude of the overall error. This justifies the empirical rule stated in Chapter 4, that in multiplication or division, the number of significant figures retained in the answer should be that in the least accurate quantity entering the calculations.

EXERCISES

1. A student determines the density of a liquid by weighing a sample from a 20 ml pipette, believed to be accurate to ± 0.01 ml. The mass, found by taking the difference between two weighings, is 16.820 g. The estimated error of each weighing is ± 0.002 g.

    a. Estimate the error in the density, taking the errors in m and V to be indeterminate.

    b. If the errors in each weighing were *known* to be 0.002 g, and that in V was *known* to be 0.01 ml, what would be the error in the density?

2. Use the relationships of differential calculus (Section 9.5) to derive the rules given for determinate errors in Table 11.6 (see Example 9.10 and the exercises at the end of Section 9.5).

PROBLEMS

11.1 A student determines the percentage of chlorine in a compound by titrating a weighed sample with $AgNO_3$. His measurements are:

mass beaker	A	volume $AgNO_3$	V
mass beaker + sample	B	conc. $AgNO_3$	M

$$\text{percentage of Cl} = \frac{100 \times V \times M \times 35.5}{(B - A)}$$

What effect will each of the following determinate errors have on the value he reports for the percentage of chlorine?

  a. He "overshoots" the end point of the titration, adding excess $AgNO_3$.
  b. The measured mass of the beaker is one gram greater than the true value.
  c. The silver nitrate solution is diluted with water before titrating.
  d. Before titration, the sample is dissolved in water containing Cl^- ions.

11.2   A student analyzing a sample for the percentage of bromine makes four trials with the following results:

$$36.0, 36.3, 35.8, \text{ and } 36.3$$

Calculate:

  a. The arithmetic mean.
  b. The deviation and percent deviation of each trial.
  c. The average deviation.
  d. The standard deviation.

11.3   The atomic weight of osmium was reported to be 190.2, based on the mean of 6 determinations with a standard deviation of 0.1.

  a. Using Table 11.3, estimate the probability that, if another determination were made, the value found would be:
    (1) within 0.1 of 190.2 (i.e., 190.1–190.3)
    (2) within 0.2 of 190.2 (i.e., 190.0–190.4)
  b. Using Table 11.4, calculate the reliability of the mean at the 50 per cent, 90 per cent, and 99 per cent confidence levels.

11.4   A student uses four different voltmeters to measure the voltage of a zinc-copper cell. He obtains the following results:

$$1.10, 1.21, 1.08, \text{ and } 1.12$$

Apply the Q test to decide whether any of these measurements should be neglected, indicating a faulty voltmeter.

11.5   A class of 10 students determines the percentage of aluminum in an Al-Zn alloy. The results are as follows:

$$60, 62, 39, 64, 58, 56, 61, 60, 69, \text{ and } 62$$

The instructor wishes to establish a reliable mean from these results. Apply the Q test to decide which, if any, of the results should be rejected.

11.6 A class of 10 students reports the following results for the percentage of chlorine in a sample:

$$20.30, 20.18, 19.98, 20.06, 20.22,$$

$$20.08, 20.02, 20.14, 20.07, \text{ and } 20.00$$

Calculate the arithmetic mean and the reliability limits for the mean at the 90 per cent confidence level. What, precisely, do these limits mean? Under what conditions could we say, with 90 per cent confidence, that the true percentage of chlorine lies within these limits?

11.7 The molecular weight of a gas can be calculated from the equation:

$$M = \frac{g \times R \times T}{P \times V}$$

where g = mass of gas, R = gas constant = 82.1 ml atm/mole°K, T = temperature in °K, P = pressure in atmospheres = pressure in mm Hg/760, V = volume in ml.

A student obtains the following values, with the indicated indeterminate errors:

$$T = 298.5 \pm 0.1°\text{K}$$
$$P = 715 \pm 1 \text{ mm Hg}$$
$$V = 260 \pm 1 \text{ ml}$$
$$g = 1.545 \text{ g} \pm 1 \text{ mg}$$

Calculate the molecular weight of the gas and estimate the error.

CHAPTER 12

INTRODUCTION TO COMPUTERS

Most of the mathematical problems encountered by students in chemistry can be solved in a relatively short time. A few moments with a slide rule, a calculating machine, logarithms, or even long division or multiplication by hand are usually all that are required to make a calculation. Nearly all of the problems we have discussed in this book, involving stoichiometry, equilibrium, and pH, even the calculus applications, require for the most part only the ubiquitous slide rule for very adequate solutions.

About the only problem we mentioned which possibly might not fall in the category which can be handled by very short calculations is that of the least squares fitting of a mathematical function to a set of experimentally obtained data points. If, for example, we obtained, in an experiment in which we measured the density of a liquid as a function of temperature, 30 data points between 0°C and 150°C, and we wished to use the least squares method to get the best fit of the function:

$$d = A + Bt$$

to those points, it would require about 70 calculations on the slide rule, and we would have to obtain 4 sums of 30 terms each before we could determine the best values of A and B in that function. In such a case, although each calculation is really easy to make, the number of such calculations is so large as to make the overall problem extremely tedious and a terrible bore. In addition, humans being as they are, the odds are that somewhere along the line we would be very likely to make a mistake, and would have to do the whole job again, and maybe even a third time.

In chemistry, there are a reasonably large number of physically important problems in which a tremendous number of simple calculations are required to produce a desired solution. In addition to the least squares fit problem, we might mention crystal structure determination by x-ray diffraction, energy levels of atoms and molecules as determined by quantum mechanics, and interatomic distances and forces as found from experiments in molecular spectroscopy. In each case, we seek relations similar to those in the least squares problem to obtain a best fit between a model we create and some experimental or theoretical conditions that that model must satisfy.

Twenty years ago, these lengthy problems were in many instances essentially impossible. In simpler cases, graduate students spent six months to a year (before mechanical calculating machines) to solve one problem and, finally, earn their Ph.D.'s. (Actually, doing calculations of these kinds wasn't all that uninteresting, and certainly did ensure that *this* graduate student, at least, knew what he was trying to do.)

During these past twenty years, a real change has been made in the way in which such lengthy calculations are carried out. Machines called computers have gradually been developed which can store large amounts of data and then perform calculations on these data according to directions which are also stored. In the twinkling of an eye, these devices can automatically carry out a multitude of arithmetic and logical processes without error and print out the results as fast as they are obtained. These days, a graduate student, with the aid of a computer, needs only a few minutes to carry out the calculations that previously took a year. Modern computers are truly impressive in their capabilities, and in the remainder of this chapter we will consider briefly how they developed historically, how they work, and how we communicate with them.

12.1   A SHORT HISTORY OF THE COMPUTER

Mechanical calculating machines have been in existence, albeit in very small numbers, for about three centuries. As a young man, Blaise Pascal, a famous French mathematician, invented and had built the first adding machine, with wheels and gears similar to those in contemporary desk adders. Leibniz, a German philosopher and mathematician, one of the discoverers of calculus, invented a machine in about 1670 which could add, subtract, multiply, divide, and take square roots. (Incidentally, Baron Gottfried Wilhelm von Leibniz must have been a truly interesting person. He was active in religion, politics, and philosophy, as well as being a first-rate mathematician and scientist.)

In the early nineteenth century, a British mathematician, Charles Babbage, added significantly to Leibniz' work by developing many of the basic ideas behind the modern computer, including even the punched card method of introducing data. Babbage spent a large part of his life designing and attempting to build his machine, which he called an Analytical Engine. He never was able to complete what would have been the first computer, possibly because the technology of the time was inadequate and possibly because Babbage was a better theoretician than he was an experimentalist. In any event, after investing about £17,000, a lot of money in those days, and getting nothing tangible in return, the British government bowed out of the scheme. Babbage died an unknown and bitter man, convinced that governments did not properly appreciate the value of science. It was only after the development of modern computers that mathematicians discovered Babbage's work and the fact that his ideas on computers were basically the same as those used today.

In the interval between 1820 and 1920, the main developments in the area of computers were the production of mechanical accounting machines and the recognition of the value of punched cards for storing data. In 1925, Mark I, the first modern mechanical computer, was designed by Dr. Vannevar Bush and built at MIT. The first automatic computer was built in 1944 by the International Business Machine Company, IBM, which had begun manufacturing accounting machines in about 1900, and had grown as needs for data handling equipment had increased. IBM is now the largest manufacturer of computers and is one of the largest corporations in the world.

The Mark I did not rely on gears, as did all the earlier machines, but was run by electrically operated mechanical switches. It was superseded in 1945 by the ENIAC, an all-electronic computer, built at the Aberdeen Proving Ground in Maryland for the purpose of ballistics calculations in World War II. EDSAC, built at Cambridge University in 1945, incorporated stored directions into the computer, and furnished the basic design for several independently built and commercial computers. The tremendous speed with which electronic computers could calculate, 5000 operations per second for the ENIAC as compared to about 3 per second for the mechanical Mark I, quickly convinced scientists that the future of computers lay in electronic rather than mechanical systems.

An important further improvement was the introduction of memory into a computer, first accomplished in the Whirlwind, built at MIT in about 1950. This invention really made the modern computer possible, and made it clear that computers would have many uses in modern society in addition to facilitating scientific calculations.

Since 1950, the expansion of the computer industry has been tremendous. The invention of the transistor and the chip, in which entire micro-electronic circuits are built in a piece of silicone semiconductor about as big as a grain of rice, made possible smaller, faster, cooler, and cheaper computers. Manufactured commercially by several companies in many sizes and prices, computers have become common equipment in industry, government, and academic life. At present, virtually every university and many small colleges in the United States have one or more computers. The IBM 1130 computer, probably the most common computer in the smaller installations, is about as large as two large desks, can make over 100,000 additions in one second, has an active memory for about 8000 numbers and a possible storage memory for millions of numbers, will take input data and product output data from or to punched cards, typewriter, or other faster methods, and in its basic form costs about $50,000, which, considering its capabilities, is inexpensive indeed.

Interestingly enough, the main uses of computers have long ceased to be in long scientific calculations of the sort we mentioned earlier. Although modern computers can and do perform many, many calculations on the basis of a relatively small amount of data, it is often more necessary to perform only a few simple calculations on a multitude of data. Many important problems of this latter kind arise in the business world. As a specific example, let us consider the booking of tickets by an airline. In operating an airline, it is important for employees at the ticket desk to be able to know current booking for every one of their flights from perhaps a month before the flight occurs until the actual time of the flight. On the basis of this information, tickets can be sold which guarantee the purchaser a seat on that flight, or if space is sold out, put the passenger on a standby list. This information should be available to all ticket offices in the country, or even in the world, on a moment's notice. A single large computer, connected by teletype machines to each of the many agents, can store the needed data, including passengers' names in many cases, can automatically give immediate statements on seat availability for all of the airline's flights, and can add or delete seats and passengers' names as tickets are bought or turned back for credit. Without this type of bookkeeping, which is really very simple for the computer, but which may involve millions of entries a month, handling of tickets by airlines would be extremely difficult and would give rise to all sorts of errors and unhappy customers.

Very similar situations arise in banks in connection with check processing, in department and mail order houses in connection with inventories, in insurance companies with billing, and in scientific organizations with the matter of storage and retrieval of scientific data. In these applications, computers offer an efficient method for handling tremendous amounts of information of all sorts; storing data, processing it in simple ways, and producing useful printed results are all readily done by the computer.

Another large application of computers that was probably not foreseen in the early days is in the area of process automation. Computers can be used to operate precision machinery in the manufacture of all sorts of simple or complex machine parts, and can take on the major control role in large chemical works. In many chemical plants, computers receive, interpret, and act on information from the many analytical sensors on the production line which monitor reagent concentrations, temperature, and pressure. Using this information, the computers activate valves, start or turn off pumps, and adjust other controls on the line in such a way as to keep the product at acceptable quality levels, often even in the face of sudden unexpected failures in the system. It has been the experience of chemical producers that computer control of chemical plants has actually improved overall product quality and decreased the amount of material that has to be rejected.

As you read this, you may easily conclude that computers can think. Indeed, they are now performing many jobs which ten or twenty years ago would have been done by men with moderate or high skill. Computers can take in information, make decisions on the basis of this information, and then proceed to take the proper action, much as a man would do, except that computers can do it more rapidly, do not get confused if too much data comes in at once, and do not get tired. However, it cannot be said that computers really think. Like a dependable and docile employee, a computer does what it is told. However, before a computer can take over in a chemical plant, a man or men must furnish the computer with information about the proper values of all the process variables with which it will deal, what the proper action is to bring those variables to the correct values should they go astray, and what to do if all hell breaks loose. This information is stored permanently in the memory of the computer and is changed as unforeseen eventualities occur. After a period of process shakedown, the computer can be given control, and, monitored by only a small number of men, it will operate the plant.

The importance of computers in our society, for good or bad, is real, and will probably increase as our technological development increases. At this point, the computer industry is booming. There are about 5000 graduate engineers, mostly electrical and very often with advanced degrees, who are working in about a half dozen large companies on the design and development of computers. About 60,000 skilled men and women are employed in the manufacture of computers, which requires careful wiring, assembling, and testing, similar to that used in making television sets. About 50,000 operators run computers in the many installations in companies, research laboratories, and universities. Nearly 500,000 people are currently computer programmers; these workers have the job of writing the lists of instructions which make it possible for the computer to handle given problems. It would only be fair to say that in terms of overall effect on our society, the computer has been the most important technological development in the United States in the past 20 years.

12.2 HOW COMPUTERS WORK

Having noted some factors in the development of the computer, we will examine in this section some of the components of computers and how they interact in actual operation.

As you might expect, the details of computer design are not terribly simple, but it is possible to discuss many of the general principles rather easily, and, hopefully, illuminatingly. In all discussion which follows, we shall use the characteristics of the very common, relatively small, IBM 1130 computer in all our illustrative comments.

For the most part, the modern computer is an electronic device. it does contain some mechanical units, used mainly in connection with input and output, but the handling of information within the computer is primarily by electrical means. At the heart of the computer is its memory, which consists of two parts, which differ in the speed at which they can be made to take in or furnish information. The fast access memory, often called the core memory, consists of grids of tiny doughnut-shaped magnets, or cores, supported on a net of wires. These grids are stacked one over the other, with the magnetic cores in corresponding positions in each grid being connected into larger units, which are called words. Each magnet may exist in one of two states, depending on its direction of magnetization; these states are assigned the numerals 1 or 0. Each magnet stores a unit of information as a 1 or a 0; each unit is called a bit. In the IBM 1130, a word in memory consists of 16 bits; the core memory contains, on the average, 8192 words, and can be assembled, amazingly enough, so that it fits in a box $3'' \times 5'' \times 5''$.

The other part of the memory in a computer is a disk, much like a phonograph record, which is coated on both sides with a magnetic material very similar to that used on recording tape. Such a disk can store over 500,000 16-bit words. When the computer is operating, the disk spins at about 1500 rpm and is scanned for loading and unloading information by access arms that move radially across its top and bottom. Whereas access time to core memory is of the order of about 4×10^{-6} seconds, it may take up to $1/2$ second for access to the information on the disk, so the use of the disk is mainly for storage of large amounts of information, which may be moved from time to time to or from core memory, in which all the rapid numerical operations of the computer occur.

All the information in a computer memory must be stored as 16-bit words, each of which consists of some arrangement of 1's and 0's. For example, three words in the computer memory might be:

0101 1110 0010 1111   1100 0000 0101 1001   1100 0000 1101 1001

In order for the computer to produce useful results, it is necessary that it be able to interpret words of the above sort in a meaningful way. There are several kinds of interpretations which are necessary for data processing; these may be numerical, alphabetical, or instructional in nature. That is, if the computer is to operate, it must essentially be able to store numbers of many possible kinds, letters, and other symbols, and it must also be able to store instructions for arithmetical processing and other operations it may be called upon to perform on numerical or alphabetical data. Since the computer can actually only store 16-bit words, it is necessary that there be a code by which such words can be interpreted to be numbers, letters, or instructions.

The code by which numbers are stored in the memory is implied by the 0 or 1 interpretation given to the states of the magnetic cores. If we decide to use the so-called binary system for numbers, which is based on 2 rather than on 10, as is our ordinary decimal system, then it is very easy to interpret any number expressed in binary bits in an unambiguous, useful way. Since you may not be familiar with the binary system, we might first write a few integers in both the decimal and binary systems:

Decimal	0	1	2	3	4	5	6	7	8	9	10
Binary	0	1	10	11	100	101	110	111	1000	1001	1010

In the decimal system, we interpret the number 258 as follows:

$$258 = 2 \times 10^2 + 5 \times 10^1 + 8 \times 10^0$$

In the binary system, taking account that the base is 2 rather than 10, the interpretation is completely analogous to that above. Consider the binary number 101101. Here we figure it out as:

$$101101 = 1 \times 2^5 + 0 \times 2^4 + 1 \times 2^3 + 1 \times 2^2 + 0 \times 2^1 + 1 \times 2^0$$

which can be expressed in normal decimal form as $32 + 8 + 4 + 1$, or 45. Any decimal integer can be expressed in binary notation with little difficulty, once you get the idea of the system. It is possible to add, subtract, multiply, or divide binary numbers very readily; indeed, any arithmetical operation can be carried out in the binary system. Since the nature of the memory bit in the computer lends itself naturally to the binary system, that system is used universally in computers. Positive and negative integers are denoted by the first bit in the 16-bit word; the number is positive if the bit has the value 0 and negative if it is 1. The remaining 15 bits can be used to express any integer up to 32,767, which in binary is 0111 1111 1111 1111. Numbers involving decimal points, so-called real numbers, are also expressed in the binary system in the computer, but the procedure used is a bit more complicated than the one we have discussed for integers, and we won't consider it further here.

To express letters in terms of the 16-bit words, there is no obvious code, such as we have with binary numbers, and the system is simply defined such that certain arrangements of bits mean, under certain conditions, letters or other useful symbols. For example, the first few letters in the alphabet are coded in the following way:

Integer Equivalents

A	1100 0001 0100 0000	−16,064
B	1100 0010 0100 0000	−15,808
C	1100 0011 0100 0000	−15,552
D	1100 0100 0100 0000	−15,296

In the right column are listed the integers which might also be represented by the word. In actual operation, the computer is instructed as to whether to treat a given word as a number or letter.

The operating instructions which allow the computer to calculate, make logical decisions, produce output, and the like, are also coded in much the same way as the letters. Below we have listed two typical instructions:

ADD   1000 0000 0000 0010   LOAD   1100 0000 0010 0000

Given the first instruction, the computer will add a number, found at an address in core memory given by the last eight bits in the instruction word, to a number in an active register, called the accumulator. The LOAD command will cause the computer to move a number from a location, again included in the last eight bits in the instruction word, into the accumulator. There are many such instruction words, and each must be stored in the computer memory so that at the proper time it can activate a chain of events which will lead to the instruction being carried out.

In the early days of computers, if one wished to do a problem on a computer, he had to write out a completely detailed list of all the instructions which had to be followed in the course of handling the problem. These instructions had to be given in terms of the 16-bit code words which represented them; this operation would be called programming the computer in machine language. Programming in machine language was, as you may well imagine, extremely difficult, slow, tedious, and subject to many errors. It was clear that if computers were to become really practical, it would be necessary to be able to instruct computers with programs written in a simpler language, more like a combination of English and algebra, hopefully expressed in common words and equations that one could easily write and recognize.

Progress in the development of simple, useful language for programming computers has been excellent during the ten or fifteen years that this kind of work has been going on. Several languages have been devised, the more common of which are known as FORTRAN, COBOL, BASIC, and ALGOL. Probably the most common of these languages is FORTRAN, from FORmula TRANslation. (In the next section we will show how a simple computer program is written in the FORTRAN language.)

In order for the computer to operate, it is essential that it be able to receive the instructions in a program. This can be accomplished in various ways, but the most common means is to use a deck of punched cards. Using a device called a key punch, which looks much like a typewriter, one can type FORTRAN statements on the cards, one statement to a card. As the typing proceeds, holes are punched into the cards according to a definite code. In Figure 12.1 are shown the punch marks made for the numbers, letters, and symbols which are used in FORTRAN. You'll notice that at the top of the card the typewriter has also printed out what was typed, so that the operator can see if he has done it properly. When a FORTRAN program has been written, the next step is to prepare the punched card deck for that program. Depending on the program, the deck may include ten or twenty cards, or a thousand or more.

When the card deck has been punched and checked to see that it is correct, it can be used as input to the computer via another device called a card reader. In this machine, the cards are examined one at a time by an optical fiber method, which senses the hole positions and reports by electrical leads to the computer the contents of each card. A card reader can examine about 500 cards a minute, which is pretty fast, but perhaps not fast enough if a program is very long. In such cases, the program, on punched cards, is fed through the card reader and put on a reel of magnetic tape. Valuable computer time can be saved by then using the magnetic tape to actually furnish the program to the computer.

We are now in a position to examine what actually happens in a computer as it executes a FORTRAN program. On being informed of the nature of the program by the card reader, the computer loads the FORTRAN dictionary, or compiler, into the core memory from the disk, on which it is permanently stored. The FORTRAN compiler is itself a set of instructions which allows the computer to translate a program written in FORTRAN into machine language. The actual FORTRAN program, which is stored in core memory as the cards are read, is then translated by the compiler. Once the program has been expressed in machine language, there is no further need for the FORTRAN compiler, and it is removed from core memory and returned to the disk. Its space in core memory, which is quite limited, is then taken by the machine language program.

In the execution of the program, the computer is run by a control unit, which automatically makes the computer move from one instruction to the next as each instruction is carried out. Data, as necessary, are moved from core memory to the accumulator register. Calculations, such as addition or multiplication, are accomplished in the accumulator by an arithmetic unit, which responds in a prescribed way to its instructions. Once obtained, the results are returned to core memory, where they are stored and are available for printout, either on a typewriter mounted on the computer console or on a special high-speed printer.

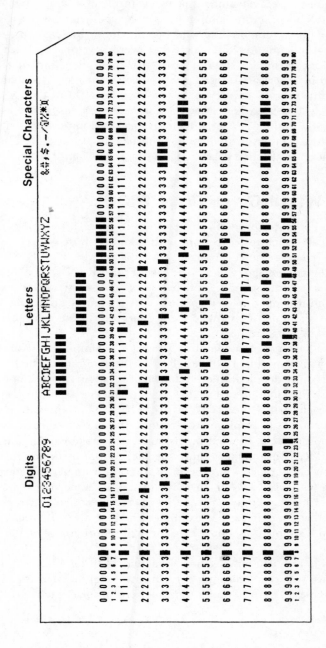

FIGURE 12.1. IBM card showing how numerals, letters, and symbols are punched for program input.

One of the real advantages of using problem-oriented languages like FORTRAN to program computers is that these languages can be employed on many computers, whereas machine language programs are limited to the computer for which they were developed. Since scientists tend to work on similar sorts of problems, they often find that programs developed in other laboratories are useful for their own problems. It is now possible to draw on very extensive libraries of FORTRAN programs, and so to deal with a wide variety of problems very easily and without the necessity of actually writing many programs. It is important that someone using a library program understand thoroughly what it does, but this is usually much easier than developing the program from scratch.

12.3   WRITING *FORTRAN* PROGRAMS

Most computer programs are written in problem-oriented languages, since they are much easier to work with than are those written in machine language or in assembler language, which is intermediate between actual machine language and a language like FORTRAN. The FORTRAN language is general enough to make use of most of the computer's capabilities, and it is only in very special cases that it is profitable to bypass FORTRAN and use a language more "natural" to the computer and less natural to the programmer.

As we have noted, a computer program is simply a list of directions by which a given problem can be solved by the computer. The FORTRAN program is quite detailed, since even in such a language, the computer does not acquire imagination or insight. It must be told specifically what the names of each variable and constant are, where to find the data it is to use, in what form it is to print out the results, the nature of each step in the computation process, to what precision it is to make calculations, and so on. FORTRAN grammar is quite specific, so that a comma missing or a parenthesis out of place will prevent a program from getting started. It is easy, even with FORTRAN, to make mistakes. One nice feature of the computer, though, is that it can be instructed to notice and report errors by typing them out after it has examined the program. It can't correct an error, even a simple one, but it can and will report many of the common ones.

In order that you may see how an actual FORTRAN program is constructed, we will carry through the entire development in a simple case. Let us assume we have 10 quadratic equations to solve, each of the form:

$$Ax^2 + Bx + C = 0$$

We would like to write out a program which would print out, for each equation, the coefficients A, B, and C, and the roots x_1 and x_2 for that equation. If the roots are not real numbers, as may be the case, we wish to have the computer so inform us, but not to determine their complex values.

Our FORTRAN program will consist of a series of statements, informing the computer how to proceed. The first thing we must do is figure out how *we* would go about it if we ourselves were to find the roots, say on a slide rule with a piece of paper. A simple procedure might be:

1. Read A, B, and C, the coefficients of x^2, x, and the constant, respectively, for the first equation.

2. Recognize the fact that the quadratic formula will allow us to find the roots of the equation. We remember that the formula is:

$$x = \frac{-B \pm (B^2 - 4AC)^{1/2}}{2A}$$

3. On a slide rule, we find both B^2 and $4AC$ and take their difference on a piece of paper. If $B^2 - 4AC$ is negative, we cannot take the square root and get a real number, so under those circumstances, we might write alongside that equation that the roots are not real.

4. If $B^2 - 4AC$ is positive, we might assign it the symbol D and find its square root, $D^{1/2}$. We also find the value of $2A$. We then solve for the first root:

$$x_1 = \frac{-B + D^{1/2}}{2A}$$

and then for the value of the second root:

$$x_2 = \frac{-B - D^{1/2}}{2A}$$

5. Having solved the first equation, we go on and read the coefficients in the second equation and solve it as we did the first. We continue in this way until we have completed all the 10 equations, at which time we decide that the job is done.

The FORTRAN program for this problem can be written to direct the computer to do essentially what we have described. The first two cards in the program tell the computer that it's going to have to go to work, and that it is to work in the FORTRAN language. There is a third card, which tells it that input and output information will involve punched cards and the console typewriter. The statements on these three cards would simply be

```
//   JOB
//   FOR
 *   IOCS (TYPEWRITER, CARD)
```

The next card tells the computer that the first equation processed is called number 1:

$$N = 1$$

Now we instruct the computer to read the coefficients in the quadratic equation from the next data card. The statement is:

$$5 \ \text{READ} \ (2,1) \ A, \ B, \ C$$

The number 5 before the statement serves to distinguish it, since we will be interested in returning later to that statement to read another data card. The numbers (2, 1) tell the computer to read a punched card, which is an input device with the code number 2, on which the coefficients A, B, and C are punched according to a layout described in a statement numbered 1. The next card specifies that layout:

$$1 \quad \text{FORMAT} \ (3F6.2)$$

This statement tells the computer that the data card will have three numbers on it, not necessarily integers, and that the first number will be somewhere in the first six columns on the card, the second will be in the next six columns, and the third will be in the next six columns. The computer will read these numbers in order and will take them to be A, B and C, in that order. You can see that this kind of instruction is necessary, since the computer would otherwise not know where to find the numbers on the card.

Having read the first data card and put the coefficients into core memory, the computer can begin to solve the equation. The next order is:

$$D = B * B - 4 * A * C$$

This expression rather looks like an equation, and it is. It tells the computer to define a quantity called D and to assign a value to D which is equal to $B^2 - 4AC$. In FORTRAN, the multiplication symbol is *, and it must always be included in such an expression if multiplication is to be done.

The computer must now check to see if the equation has real roots. This is done by the following instruction:

$$\text{IF} \ (D) \quad 15, 20, 20$$

This rather odd-looking expression orders a logic step, rather than a strictly arithmetic one. It means, "Examine the sign of D. If D is negative, go to statement number 15. If D is either zero or is positive, go to statement number 20." You can see that if the first condition is true, the equation does not have real roots, and we wish to have the computer indicate that. If the second or third possibility is true, then real roots exist for the equation, and we must proceed to find them. Statement 15 will be an output order:

$$15 \quad \text{WRITE} \quad (1,2) \ N, A, B, C,$$

As you might now surmise, this order would cause the computer to type out on the computer console printer the number of the equation and the values of A, B, and C, according to the format given in statement 2. Statement 2 might be:

$$2 \quad \text{FORMAT} \quad (/, 3X, I2, 4X, F5.1, 1X, F5.1, 1X, F5.1, 5X,$$

$$\text{'ROOTS ARE NOT REAL'})$$

This layout statement is a bit more complicated than the previous one, but it is really not too bad. The / means turn the typewriter roll one space, 3X means hit the space bar three times, and I2 means the first number of the card will be an integer of up to 2 digits; then follow the values of A, B, and C, each of which can be a noninteger up to 5 spaces wide, including any negative signs and decimal points, and then follows the printed statement that the equation does not have real roots. If the roots are not real, once that has been reported, we can direct the computer to start work on the next equation by the statement:

$$\text{GO TO} \quad 25$$

This order will cause the computer to proceed immediately to statement 25, which we will consider in a moment.

Having taken care of the case where there are no real roots, we return to statement 20, in which we evaluate the first root:

$$20 \ X1 \ = \ (-B + \text{SQRT} \ (D)) \, / \, (2 * A)$$

followed by a card on which we have the order to find the second root:

$$X2 \ = \ (-B - \text{SQRT} \ (D)) \, / \, (2 * A)$$

Built into the computer is the possibility of doing all the common arithmetic operations. This order would result in the computer's calculating the value of $-B + (D)^{1/2}$, dividing by $2A$, and calling this quantity X1. Within the disk memory are instructions for calculating many useful quantities, in addition to instructions for adding, subtracting, and the like.

These include the trigonometric functions, logarithms, and other functions, such as square root. When the computer reads a card on which SQRT appears, it automatically will load into its core memory the instructions for taking square roots, so that when the need arises during the calculations, that operation will be possible. Having had the computer find the roots of the equation, we order that they be printed out:

$$\text{WRITE} \quad (1, 3) \; N, A, B, C, \text{X1}, \text{X2}$$

We arrange that the results for equations with roots line up with those which don't have roots by FORMAT statement 3:

3 FORMAT $(/, 3X, I2, 4X, F5.1, 1X, F5.1, 1X, F5.1, 5X, F7.3, 3X, F7.3)$

Here, the directions are the same as in statement 2, except that we have included space for the two roots.

The computer will follow the instructions in the program in order unless directed, as by the IF statement, to do otherwise. At this point in the problem, the computer will have printed out results for the first quadratic equation, and we must direct it to go back and start on the next equation. This procedure is to be repeated until the ten equations have been dealt with. We can get the computer to carry out these orders by two simple statements:

$$25 \quad N = N + 1$$
$$\text{IF} \; (N - 10) \; 5, 5, 30$$

Note that the computer will get to statement 25 whether the equation has real roots or not. This statement tells the computer to replace the value it has stored for N by $N + 1$. At the end of the calculation for the first equation, N equals 1, since we so defined it early in the program; after passing this statement, the value of N which is stored will be 2. In the IF statement, the computer examines the value of $N - 10$. If N is less than 10, the computer will return to statement 5 and read the coefficients in the next equation and then proceed to work with them. If N is equal to 10, as it would be just before the 10th equation is to be done, we still wish to return to statement 5 and treat the 10th equation. When the computer has completed processing equation 10, it will assign 11 to the value of N, and the value of $N - 10$ becomes positive for the first time. It is then that we wish to have the computer stop working and we issue the order in statement 30, which is where the computer will go if $N - 10$ is positive:

$$30 \quad \text{CALL EXIT}$$

followed by two terminating statements:

END

// XEQ

In processing the program, the computer will read all the instruction cards in order, but will not read any data cards until it has ascertained that the instructions make sense to it and that it is ready to handle the data. If, by some chance, there is an error in the program, such as a missing statement number, or a call for printout of a quantity we failed to define, then the computer will stop after reading the // XEQ card and report the kind of error that is present. It will not read any data cards until the error has been rectified. When all errors have been removed, which will of course require making some change in the deck of punched cards and passing them through the card reader again, the actual computation of the roots will proceed automatically after the XEQ card has been read. The data cards immediately follow the XEQ card in the card deck, and are read only after the computer has examined the program and found it correct.

Often it is convenient to have the computer actually type out the whole FORTRAN program as written. This can be accomplished by inserting a card just after the IOCS card on which we have the instruction:

* LIST SOURCE PROGRAM

Another useful operation would be to have a title over the output from the program, along with column headings indicating the quantities that appear in the various columns in the table; this could be accomplished by a WRITE statement followed by a FORMAT statement, both placed just after the LIST SOURCE PROGRAM card:

WRITE   (1, 6)

6   FORMAT (10X, 'QUADRATIC EQUATIONS PROBLEM',//,

'EQN. NO.', 3X, 'A', 5X,

'B', 5X, 'C', 8X, 'X1', 7X, 'X2')

In Figure 12.2 we have shown the computer output for this program for a sample set of data, along with the source program. Writing the program, punching the cards, and obtaining the results took roughly an hour for someone reasonably familiar with the computer. Actual computer running time was on the order of two minutes, of which the time required to read

```
// JOB

// FOR

*LIST SOURCE PROGRAM
*IOCS(TYPEWRITER,CARD)
      WRITE(1,6)
6     FORMAT(10X,'QUADRATIC  EQUATIONS PROBLEM',//,'EQN. NO.',3X,'A',5X,
     1'B',5X,'C',8X,'X1',7X,'X2')
      N=1
5     READ(2,1)A,B,C
1     FORMAT(3F6.2)
      D=B*B-4*A*C
      IF(D)15,20,20
15    WRITE(1,2)N,A,B,C
2     FORMAT(/,3X,I2,4X,F5.1,1X,F5.1,1X,F5.1,5X,'ROOTS ARE NOT REAL')
      GO TO 25
20    X1=(-B+SQRT(D))/(2*A)
      X2=(-B-SQRT(D))/(2*A)
      WRITE(1,3)N,A,B,C,X1,X2
3     FORMAT(/,3X,I2,4X,F5.1,1X,F5.1,1X,F5.1,5X,F7.3,3X,F7.3)
25    N=N+1
      IF(N-10)5,5,30
30    CALL EXIT
      END

FEATURES SUPPORTED
  IOCS

CORE REQUIREMENTS FOR
  COMMON      0  VARIABLES     18  PROGRAM     232

END OF COMPILATION

// XEQ
          QUADRATIC  EQUATIONS PROBLEM
```

EQN. NO.	A	B	C	X1	X2
1	1.0	-5.0	6.0	3.000	2.000
2	1.0	2.0	0.0	0.000	-2.000
3	4.0	6.0	-3.0	0.395	-1.895
4	2.0	2.0	5.0	ROOTS ARE NOT REAL	
5	9.0	10.0	0.0	0.000	-1.111
6	4.0	1.0	5.0	ROOTS ARE NOT REAL	
7	1.0	0.0	-4.0	2.000	-2.000
8	6.0	0.0	1.0	ROOTS ARE NOT REAL	
9	1.0	6.0	9.0	-3.000	-3.000
10	1.0	0.0	0.0	0.000	0.000

FIGURE 12.2. Computer printout on quadratic equation problem.

the program cards, load the FORTRAN compiler, and print everything but the actual results was about one and one-half minutes.

Computer programming is an exacting, logical activity, found by many to be very fascinating. To learn enough FORTRAN programming to do most routine calculations takes a relatively small amount of time, perhaps a week or so to become familiar with the language and some of the most common pitfalls. The use of the computer is becoming more important for scientists and for science students, and in a few years it is entirely possible that college students will become as familiar with the campus computer as they are now with the slide rule. If you are interested in exploring computers and their uses further, you might take a stroll down to the campus computer center and begin talking with some of the devotees down there. Who knows, you may never want to leave!

SUGGESTED REFERENCES

General

Richardson, M.: Fundamentals of Mathematics, 3rd ed. Macmillan, 1966.
May, K. O.: Elementary Analysis. John Wiley and Sons, Inc., 1952.
Beiser, A.: Essential Math for the Sciences. McGraw-Hill, 1969.
Greenberg, D. A.: Mathematics for Introductory Science Courses. W. A. Benjamin, 1965.

The Slide Rule (Chapter 3)

Saffold, R., and Smalley, A.: The Slide Rule. Doubleday and Co., Inc., 1962.
Breckinridge, W. E.: The Polyphase Slide Rule. Keuffel and Esser, 1944.

Functional Relationships. Graphical Analysis (Chapter 7).

Ellerby, G.: Graphs and Calculus. Pergamon Press, 1964.
Davis, D. S.: Empirical Equations and Nomography. McGraw-Hill, 1943.

Space Geometry and Trigonometry (Chapter 8)

Federer, H., and Jónsson, B.: Analytic Geometry and Calculus. Ronald Press, 1961.
Bruce, W. J.: Analytic Trigonometry. Pergamon Press, 1963.
Johnson, R. E., and McCoy, N. H.: Analytic Geometry. Holt, Rinehart and Winston, 1955.

Calculus (Chapters 9 and 10)

Thompson, P.: Calculus Made Easy. Macmillan, 1960.
Morse, E. E.: Calculus. Addison-Wesley, 1966.
Butler, J. N., and Bohrow, D. G.: The Calculus of Chemistry. W. A. Benjamin, 1966.

Error Analysis. Statistical Methods (Chapter 11)

Fritz, J., and Schenk, G.: Quantitative Analytical Chemistry. Allyn and Bacon, 1966 (Chap. 19).
Laitinen, H. A.: Chemical Analysis. McGraw-Hill, 1960 (Chap. 26).
Mills, F. C.: Introduction to Statistics. Holt, Rinehart and Winston, 1956.
Dixon, W. J., and Massey, F. J., Jr.: Introduction to Statistical Analysis. McGraw-Hill, 1957.

Computers (Chapter 12)

Desmonde, W. H.: Computers and Their Uses. Prentice-Hall, 1964.
Calingaert, P.: Principles of Computation. Addison-Wesley, 1965.
Laurie, E. J.: Computers and How They Work. Southwestern Publishing Co., 1963.
Dickson, T. R.: The Computer and Chemistry. W. H. Freeman, 1968.

MATHEMATICAL TABLES

A. LOGARITHMS

	0	1	2	3	4	5	6	7	8	9
1.0	.0000	.0043	.0086	.0128	.0170	.0212	.0253	.0294	.0334	.0374
1.1	.0414	.0453	.0492	.0531	.0569	.0607	.0645	.0682	.0719	.0755
1.2	.0792	.0828	.0864	.0899	.0934	.0969	.1004	.1038	.1072	.1106
1.3	.1139	.1173	.1206	.1239	.1271	.1303	.1335	.1367	.1399	.1430
1.4	.1461	.1492	.1523	.1553	.1584	.1614	.1644	.1673	.1703	.1732
1.5	.1761	.1790	.1818	.1847	.1875	.1903	.1931	.1959	.1987	.2014
1.6	.2041	.2068	.2095	.2122	.2148	.2175	.2201	.2227	.2253	.2279
1.7	.2304	.2330	.2355	.2380	.2405	.2430	.2455	.2480	.2504	.2529
1.8	.2553	.2577	.2601	.2625	.2648	.2672	.2695	.2718	.2742	.2765
1.9	.2788	.2810	.2833	.2856	.2878	.2900	.2923	.2945	.2967	.2989
2.0	.3010	.3032	.3054	.3075	.3096	.3118	.3139	.3160	.3181	.3201
2.1	.3222	.3243	.3263	.3284	.3304	.3324	.3345	.3365	.3385	.3404
2.2	.3424	.3444	.3464	.3483	.3502	.3522	.3541	.3560	.3579	.3598
2.3	.3617	.3636	.3655	.3674	.3692	.3711	.3729	.3747	.3766	.3784
2.4	.3802	.3820	.3838	.3856	.3874	.3892	.3909	.3927	.3945	.3962
2.5	.3979	.3997	.4014	.4031	.4048	.4065	.4082	.4099	.4116	.4133
2.6	.4150	.4166	.4183	.4200	.4216	.4232	.4249	.4265	.4281	.4298
2.7	.4314	.4330	.4346	.4362	.4378	.4393	.4409	.4425	.4440	.4456
2.8	.4472	.4487	.4502	.4518	.4533	.4548	.4564	.4579	.4594	.4609
2.9	.4624	.4639	.4654	.4669	.4683	.4698	.4713	.4728	.4742	.4757
3.0	.4771	.4786	.4800	.4814	.4829	.4843	.4857	.4871	.4886	.4900
3.1	.4914	.4928	.4942	.4955	.4969	.4983	.4997	.5011	.5024	.5038
3.2	.5051	.5065	.5079	.5092	.5105	.5119	.5132	.5145	.5159	.5172
3.3	.5185	.5198	.5211	.5224	.5237	.5250	.5263	.5276	.5289	.5302
3.4	.5315	.5328	.5340	.5353	.5366	.5378	.5391	.5403	.5416	.5428
3.5	.5441	.5453	.5465	.5478	.5490	.5502	.5514	.5527	.5539	.5551
3.6	.5563	.5575	.5587	.5599	.5611	.5623	.5635	.5647	.5658	.5670
3.7	.5682	.5694	.5705	.5717	.5729	.5740	.5752	.5763	.5775	.5786
3.8	.5798	.5809	.5821	.5832	.5843	.5855	.5866	.5877	.5888	.5899
3.9	.5911	.5922	.5933	.5944	.5955	.5966	.5977	.5988	.5999	.6010
4.0	.6021	.6031	.6042	.6053	.6064	.6075	.6085	.6096	.6107	.6117
4.1	.6128	.6138	.6149	.6160	.6170	.6180	.6191	.6201	.6212	.6222
4.2	.6232	.6243	.6253	.6263	.6274	.6284	.6294	.6304	.6314	.6325
4.3	.6335	.6345	.6355	.6365	.6375	.6385	.6395	.6405	.6415	.6425
4.4	.6435	.6444	.6454	.6464	.6474	.6484	.6493	.6503	.6513	.6522
4.5	.6532	.6542	.6551	.6561	.6571	.6580	.6590	.6599	.6609	.6618
4.6	.6628	.6637	.6646	.6656	.6665	.6675	.6684	.6693	.6702	.6712
4.7	.6721	.6730	.6739	.6749	.6758	.6767	.6776	.6785	.6794	.6803
4.8	.6812	.6821	.6830	.6839	.6848	.6857	.6866	.6875	.6884	.6893
4.9	.6902	.6911	.6920	.6938	.6937	.6946	.6955	.6964	.6972	.6981
5.0	.6990	.6998	.7007	.7016	.7024	.7033	.7042	.7050	.7059	.7067
5.1	.7076	.7084	.7093	.7101	.7110	.7118	.7126	.7135	.7143	.7152
5.2	.7160	.7168	.7177	.7185	.7193	.7202	.7210	.7218	.7226	.7235
5.3	.7243	.7251	.7259	.7267	.7275	.7284	.7292	.7300	.7308	.7316
5.4	.7324	.7332	.7340	.7348	.7356	.7364	.7372	.7380	.7388	.7396
5.5	.7404	.7412	.7419	.7427	.7435	.7443	.7451	.7459	.7466	.7474
5.6	.7482	.7490	.7497	.7505	.7513	.7520	.7528	.7536	.7543	.7551
5.7	.7559	.7566	.7574	.7582	.7589	.7597	.7604	.7612	.7619	.7627
5.8	.7634	.7642	.7649	.7657	.7664	.7672	.7679	.7686	.7694	.7701
5.9	.7709	.7716	.7723	.7731	.7738	.7745	.7752	.7760	.7767	.7774

	0	1	2	3	4	5	6	7	8	9
6.0	.7782	.7789	.7796	.7803	.7810	.7818	.7825	.7832	.7839	.7846
6.1	.7853	.7860	.7868	.7875	.7882	.7889	.7896	.7903	.7910	.7917
6.2	.7924	.7931	.7938	.7945	.7952	.7959	.7966	.7973	.7980	.7987
6.3	.7993	.8000	.8007	.8014	.8021	.8028	.8035	.8041	.8048	.8055
6.4	.8062	.8069	.8075	.8082	.8089	.8096	.8102	.8109	.8116	.8122
6.5	.8129	.8136	.8142	.8149	.8156	.8162	.8169	.8176	.8182	.8189
6.6	.8195	.8202	.8209	.8215	.8222	.8228	.8235	.8241	.8248	.8254
6.7	.8261	.8267	.8274	.8280	.8287	.8293	.8299	.8306	.8312	.8319
6.8	.8325	.8331	.8338	.8344	.8351	.8357	.8363	.8370	.8376	.8382
6.9	.8388	.8395	.8401	.8407	.8414	.8420	.8426	.8432	.8439	.8445
7.0	.8451	.8457	.8463	.8470	.8476	.8482	.8488	.8494	.8500	.8506
7.1	.8513	.8519	.8525	.8531	.8537	.8543	.8549	.8555	.8561	.8567
7.2	.8573	.8579	.8585	.8591	.8597	.8603	.8609	.8615	.8621	.8627
7.3	.8633	.8639	.8645	.8651	.8657	.8663	.8669	.8675	.8681	.8686
7.4	.8692	.8698	.8704	.8710	.8716	.8722	.8727	.8733	.8739	.8745
7.5	.8751	.8756	.8762	.8768	.8774	.8779	.8785	.8791	.8797	.8802
7.6	.8808	.8814	.8820	.8825	.8831	.8837	.8842	.8848	.8854	.8859
7.7	.8865	.8871	.8876	.8882	.8887	.8893	.8899	.8904	.8910	.8915
7.8	.8921	.8927	.8932	.8938	.8943	.8949	.8954	.8960	.8965	.8971
7.9	.8976	.8982	.8987	.8993	.8998	.9004	.9009	.9015	.9020	.9026
8.0	.9031	.9036	.9042	.9047	.9053	.9058	.9063	.9069	.9074	.9079
8.1	.9085	.9090	.9096	.9101	.9106	.9112	.9117	.9122	.9128	.9133
8.2	.9138	.9143	.9149	.9154	.9159	.9165	.9170	.9175	.9180	.9186
8.3	.9191	.9196	.9201	.9206	.9212	.9217	.9222	.9227	.9232	.9238
8.4	.9243	.9248	.9253	.9258	.9263	.9269	.9274	.9279	.9284	.9289
8.5	.9294	.9299	.9304	.9309	.9315	.9320	.9325	.9330	.9335	.9340
8.6	.9345	.9350	.9355	.9360	.9365	.9370	.9375	.9380	.9385	.9390
8.7	.9395	.9400	.9405	.9410	.9415	.9420	.9425	.9430	.9435	.9440
8.8	.9445	.9450	.9455	.9460	.9465	.9469	.9474	.9479	.9484	.9489
8.9	.9494	.9499	.9504	.9509	.9513	.9518	.9523	.9528	.9533	.9538
9.0	.9542	.9547	.9552	.9557	.9562	.9566	.9571	.9576	.9581	.9586
9.1	.9590	.9595	.9600	.9605	.9609	.9614	.9619	.9624	.9628	.9633
9.2	.9638	.9643	.9647	.9652	.9657	.9661	.9666	.9671	.9675	.9680
9.3	.9685	.9689	.9694	.9699	.9703	.9708	.9713	.9717	.9722	.9727
9.4	.9731	.9736	.9741	.9745	.9750	.9754	.9759	.9763	.9768	.9773
9.5	.9777	.9782	.9786	.9791	.9795	.9800	.9805	.9809	.9814	.9818
9.6	.9823	.9827	.9832	.9836	.9841	.9845	.9850	.9854	.9859	.9863
9.7	.9868	.9872	.9877	.9881	.9886	.9890	.9894	.9899	.9903	.9908
9.8	.9912	.9917	.9921	.9926	.9930	.9934	.9939	.9943	.9948	.9952
9.9	.9956	.9961	.9965	.9969	.9974	.9978	.9983	.9987	.9991	.9996

B. TRIGONOMETRIC FUNCTIONS

angle	sin	tan	cot	cos	
0°	.0000	.0000	∞	1.0000	90°
1	.0175	.0175	57.29	.9999	89
2	.0349	.0349	28.64	.9994	88
3	.0523	.0524	19.08	.9986	87
4	.0698	.0699	14.30	.9976	86
5	.0872	.0875	11.43	.9962	85
6	.1045	.1051	9.514	.9945	84
7	.1219	.1228	8.144	.9926	83
8	.1392	.1405	7.115	.9903	82
9	.1564	.1584	6.314	.9877	81
10	.1737	.1763	5.671	.9848	80
11	.1908	.1944	5.145	.9816	79
12	.2079	.2126	4.705	.9782	78
13	.2250	.2309	4.332	.9744	77
14	.2419	.2493	4.011	.9703	76
15	.2588	.2680	3.732	.9659	75
16	.2756	.2868	3.487	.9613	74
17	.2924	.3057	3.271	.9563	73
18	.3090	.3249	3.078	.9511	72
19	.3256	.3443	2.904	.9455	71
20	.3420	.3640	2.748	.9397	70
21	.3584	.3839	2.605	.9336	69
22	.3746	.4040	2.475	.9272	68
23	.3907	.4245	2.356	.9205	67
24	.4067	.4452	2.246	.9136	66
25	.4226	.4663	2.145	.9063	65
26	.4384	.4877	2.050	.8988	64
27	.4540	.5095	1.963	.8910	63
28	.4695	.5317	1.881	.8830	62
29	.4848	.5543	1.804	.8746	61
30	.5000	.5774	1.732	.8660	60
31	.5150	.6009	1.664	.8572	59
32	.5299	.6249	1.600	.8481	58
33	.5446	.6494	1.540	.8387	57
34	.5592	.6745	1.483	.8290	56
35	.5736	.7002	1.428	.8192	55
36	.5878	.7265	1.376	.8090	54
37	.6018	.7536	1.327	.7986	53
38	.6157	.7813	1.280	.7880	52
39	.6293	.8098	1.235	.7772	51
40	.6428	.8391	1.192	.7660	50
41	.6561	.8693	1.150	.7547	49
42	.6691	.9004	1.111	.7431	48
43	.6820	.9325	1.072	.7314	47
44	.6947	.9657	1.036	.7193	46
45	.7071	1.0000	1.000	.7071	45
	cos	cot	tan	sin	angle

C. FORMULAS OF CALCULUS

Derivatives

1. $\dfrac{d(a)}{dx} = 0$

2. $\dfrac{d(x^n)}{dx} = n x^{n-1}$ $\qquad\qquad$ $\dfrac{d(u^n)}{dx} = n u^{n-1} \dfrac{du}{dx}$

3. $\dfrac{d(\ln x)}{dx} = \dfrac{1}{x}$ $\qquad\qquad$ $\dfrac{d(\ln u)}{dx} = \dfrac{1}{u} \dfrac{du}{dx}$

4. $\dfrac{d(\log_{10} x)}{dx} = \dfrac{1}{2.30x}$ $\qquad\qquad$ $\dfrac{d(\log_{10} u)}{dx} = \dfrac{1}{2.30u} \dfrac{du}{dx}$

5. $\dfrac{d(e^x)}{dx} = e^x$ $\qquad\qquad$ $\dfrac{d(e^u)}{dx} = e^u \dfrac{du}{dx}$

6. $\dfrac{d(a^x)}{dx} = a^x \ln a$ $\qquad\qquad$ $\dfrac{d(a^u)}{dx} = a^u \ln a \dfrac{du}{dx}$

7. $\dfrac{d(\sin x)}{dx} = \cos x$ $\qquad\qquad$ $\dfrac{d(\sin u)}{dx} = \cos u \dfrac{du}{dx}$

8. $\dfrac{d(\cos x)}{dx} = -\sin x$ $\qquad\qquad$ $\dfrac{d(\cos u)}{dx} = -\sin u \dfrac{du}{dx}$

9. $\dfrac{d(u + v)}{dx} = \dfrac{du}{dx} + \dfrac{dv}{dx}$

10. $\dfrac{d(uv)}{dx} = u \dfrac{dv}{dx} + v \dfrac{du}{dx}$

11. $\dfrac{d(u/v)}{dx} = \dfrac{v \dfrac{du}{dx} - u \dfrac{dv}{dx}}{v^2}$

Integrals

1. $\displaystyle\int dx = x + C$

2. $\displaystyle\int x^n dx = \dfrac{x^{n+1}}{n + 1} + C \qquad (n \neq 1)$

3. $\displaystyle\int \dfrac{dx}{x} = \ln x + C$

4. $\displaystyle\int e^x dx = e^x + C$

5. $\displaystyle\int a^x dx = \dfrac{a^x}{\ln a} + C$

6. $\displaystyle\int \ln x \, dx = x \ln x - x + C$

7. $\displaystyle\int \sin x \, dx = -\cos x + C$

8. $\int \cos x \, dx = \sin x + C$

9. $\int (ax + b)^n \, dx = \dfrac{(ax + b)^{n+1}}{a(n + 1)} + C \quad (n \neq 1)$

10. $\int \dfrac{dx}{ax + b} = \dfrac{\ln(ax + b)}{a} + C$

11. $\int \dfrac{x \, dx}{ax + b} = \dfrac{x}{a} - \dfrac{b}{a^2} \ln(ax + b) + C$

12. $\int \dfrac{x \, dx}{(ax + b)^2} = \dfrac{b}{a^2(ax + b)} + \dfrac{1}{a^2} \ln(ax + b) + C$

13. $\int \dfrac{x^2 \, dx}{ax + b} = \dfrac{1}{a^3} \left[\dfrac{(ax + b)^2}{2} - 2b(ax + b) + b^2 \ln(ax + b) \right] + C$

14. $\int au \, dx = a \int u \, dx$

15. $\int (u + v) \, dx = \int u \, dx + \int v \, dx$

16. $\int u \, dv = uv - \int v \, du$

D.   STATISTICAL TABLES

1. Probability of Error, Normal Distribution (see Table 11.3, p. 173)

Range	Probability of smaller error	Probability of larger error
-0.674σ to 0.674σ	0.50	0.50
$-\sigma$ to σ	0.683	0.317
-1.282σ to 1.282σ	0.80	0.20
-1.645σ to 1.645σ	0.90	0.10
-1.960σ to 1.960σ	0.95	0.05
-2σ to 2σ	0.955	0.045
-2.576σ to 2.576σ	0.99	0.01
-3σ to 3σ	0.997	0.003
-3.291σ to 3.291σ	0.999	0.001

2. t Test (see Table 11.4, p. 176)

Confidence level

n	50%	90%	95%	99%	99.5%
2	1.000	6.314	12.706	63.657	127.320
3	0.816	2.920	4.303	9.925	14.089
4	0.765	2.353	3.182	5.841	7.453
5	0.741	2.132	2.776	4.604	5.598
6	0.727	2.015	2.571	4.032	4.773
7	0.718	1.943	2.447	3.707	4.317
8	0.711	1.895	2.365	3.500	4.029
9	0.706	1.860	2.306	3.355	3.832
10	0.703	1.833	2.262	3.250	3.690
15	0.692	1.761	2.145	2.977	3.497
20	0.688	1.729	2.093	2.861	3.174
25	0.685	1.711	2.064	2.797	3.091
∞	0.674	1.645	1.960	2.576	2.807

3. Q Test (see Table 11.5, p. 181)

n	3	4	5	6	7	8	9	10
$Q_{0.90}$	0.94	0.76	0.64	0.56	0.51	0.47	0.44	0.41

ANSWERS TO PROBLEMS AND EXERCISES

CHAPTER 1

Section 1.1

1. 1×10^3
2. 1×10^9
3. 1×10^{-6}
4. 1.622×10^4
5. 2.126×10^2
6. 1.89×10^{-1}
7. 6.18×10^0
8. 7.846×10^{-8}

Section 1.2

1. 9.30×10^{12}
2. 3.9×10^2
3. 1.05×10^{-2}
4. 3.90×10^{16}
5. 1.26×10^3
6. 5.18×10^{-3}
7. 2.0×10^{-7}
8. 3.38×10^{15}
9. 1.48×10^{-18}

Section 1.3

1. 4.67×10^{-6}
2. 1.2×10^{14}
3. 3.6×10^{-41}
4. 3.0×10^3
5. 9.2×10^2
6. 2.5×10^{-1}
7. 4.02×10^{-4}
8. 2.78×10^3
9. 2.0×10^6
10. 3.1×10^1

Section 1.4

1. 4.71×10^4
2. 5.47×10^{-2}
3. 6.20×10^4
4. 2.45×10^{-4}
5. 8.05×10^5
6. 6.37×10^{-10}
7. 9.75×10^4
8. 6.02×10^{23}

Problems

1.1 a. 6.65×10^{-24} g b. 9.3×10^{-9} cm c. 1.36×10^5 cm/sec
1.2 a. 4.00 g b. 6.15×10^{-14} erg   1.3 9.43×10^4 cal
1.4 2.99×10^{-23} g   1.5 2.5×10^{18} 1.6 2.18×10^4 cc
1.7 2.44×10^2 torr   1.8 a. 1.5×10^{-4} b. 0.95
1.9 5.3×10^{-9} cm; 2.1×10^{-8} cm; 4.8×10^{-8} cm
1.10 4.1×10^{-6} dyne   1.11 5.57×10^{-8} cm
1.12 a. 1.60×10^3 yr b. 5.84×10^5 days c. 8.41×10^8 min
1.13 a. 1.7×10^{-3} b. 4.3×10^{-2} c. 1.6×10^{-2}
1.14 a. 2×10^{-4} b. 2×10^{-3} 1.15 1.8×10^{-4}; 1.8×10^{-4}
                                                      5.6×10^{-5}; 5.6×10^{-4}
                                                      1.8×10^{-5}; 1.8×10^{-3}

CHAPTER 2

Section 2.1

1. 0.9117	3. 0.2180	5. 1.7522
2. 0.0128	4. 0.7665	6. 9.6212
7. $0.6918 - 3 = -2.3082$	8. $0.5877 - 12 = -11.4123$	

Section 2.2

1. 7.640	3. 55.40	5. 3.960×10^{-3}	7. 3.981×10^{-13}
2. 8.644	4. 41.35	6. 2.402×10^{-3}	8. 4.189×10^{7}

Section 2.3

1. 2.0149	3. -2.5853	5. 0.1974	7. 1.388×10^{16}
2. 2.1961	4. 8.8743	6. 19.2606	8. 7.208×10^{6}

Section 2.4

1. 1.796	3. -9.693	5. 2.688	7. 0.3679
2. 7.612	4. 0.4343	6. 9.33×10^{9}	

Problems

2.1 a. 6.00 b. 1.583 c. 8.52 d. -0.78

2.2 pH of solution B − pH of solution A = 2.00

2.3 2.872   2.4. a. 1×10^{-4} b. 2.5×10^{-13} c. 7.2×10^{-4} d. 10

2.5 1×10^{-15} 2.6 a. 1×10^{-8} b. 4.0 c. 11.30

2.7 a. 1, greater, less b. -5190 cal c. 4.2×10^{2}

2.8 a. -1.00 b. 5×10^{-26}

2.9 a. 9.05×10^{-3} g b. 2.3×10^{2} sec c. $\log \dfrac{2}{1} = \dfrac{k\,t_{1/2}}{2.303}$;

$$t_{1/2} = \frac{(2.303)\,(0.3010)}{k}$$

$$= 0.693/k$$

2.10 3.9×10^{-7} 2.11 a. 24 mm Hg b. $386°$K $(113°$C)

2.12 4.77×10^{-8} 2.13 a. 1 b. 0.187 c. 0.433

CHAPTER 3

Section 3.1

3. You are subtracting a distance (about 6 inches) proportional to the log of 4 (0.6021) from a distance (about 9 inches) proportional to the log of 8 (0.9031). The remainder (about 3 inches) is proportional to the log of 2 (0.3010).

4. Percent error $= \dfrac{(2.00 - 1.99)}{2.00} \times 100 = 0.5\%$

$$= \frac{(10.00 - 9.95)}{10.00} \times 100 = 0.5\%$$

Note that the percent error is approximately the same regardless of the section of the D scale you are using.

Section 3.2

1. 8.91×10^{-2} 3. 1.17×10^{11} 5. 6.81×10^{8} 7. 6.65×10^{-5}
2. 2.48×10^{1} 4. 0.217 6. 0.611

Section 3.3

1. 2.86 3. 1.26 5. 1.42×10^{1} 7. 5.46×10^{-3}
2. 3.75×10^{-5} 4. 9.33 6. 6.21×10^{3}

Section 3.4

1. 2.25×10^{2} 3. 2.11×10^{-10} 5. 6.45 7. 2.39×10^{1}
2. 2.74×10^{6} 4. 2.35 6. 1.25 8. 8.40×10^{-2}

Section 3.5

1. 0.788 3. -3.234 5. 2.90 7. 3.98×10^{-6}
2. 5.111 4. 1.035 6. 1.78×10^{1}

Problems

3.1 3.31×10^{-5} 3.2 7.18×10^{-6} 3.3 9.49×10^{14}
3.4 5.382 3.5 1.94×10^{-2} 3.6 4.46 3.7 2.24×10^{1}
3.8 2.15×10^{-66} 3.9 5.05×10^{3} 3.10 5.76×10^{3}
3.11 1.50×10^{-1} 3.12 3.87×10^{7} 3.13 3.09×10^{2}
3.14 2.51 3.15 2.69×10^{-6} 3.16 1.07×10^{-1}
3.17 2.99×10^{4} 3.18 6.05 3.19 1.6×10^{7} 3.20 8.62×10^{7}

CHAPTER 4

Section 4.1

1. 4   3. 4   5. 5   7. 3   9. 1
2. 3   4. 3   6. 3   8. 2   10. 2 or 3

Section 4.2

1. 9.49×10^{-3} 3. 1.02 5. 1.2 7. 2.4
2. 1.2×10^{2} 4. 1.01×10^{-5} 6. 1.68 8. 5.80×10^{1}

Section 4.3

1. 15.33 3. 33.7 5. 1.06×10^{2} 7. 75.2
2. 5.77 4. 908 6. 3.34×10^{-1} 8. 4.2

Section 4.4

1. 4.315465 3. 4.3155 5. 4.31 7. 4
2. 4.31547 4. 4.315 6. 4.3

Section 4.5

1. 0.2046   3. 3.339   5. 1×10^2 7. 1.6×10^3
2. 0.72   4. -3.31 6. 1.53   8. 1.25×10^{-2}

Problems

4.1 1.56 g/cc   4.2 2.2 g/cc   4.3 0.336 g;   0.159 cc
4.4 617 g   4.5 9.995 cc   4.6 a. 0.639 g b. 18.9 g
4.7 a. 1.269 g b. 35.3% C, 2.5% H, 62.21% I   4.8 a. 7.334 g, 23 g
4.9 0.230   4.10 222 g/ml   4.11 a. 10^{-4} b. 8×10^{-5}
c. 7.6×10^{-5} d. 7.62×10^{-5} 4.12 5543 cal
4.13 5.44×10^8 kcal

CHAPTER 5

Problems

5.1 a. $3.28 \text{ cm} \times \dfrac{10 \text{ mm}}{1 \text{ cm}} = 32.8 \text{ mm}$

b. $3.28 \text{ cm} \times \dfrac{1 \text{ meter}}{100 \text{ cm}} = 3.28 \times 10^{-2} \text{ meter}$

c. $3.28 \text{ cm} \times \dfrac{1 \text{ in}}{2.54 \text{ cm}} = 1.29 \text{ in}$

d. $3.28 \text{ cm} \times \dfrac{10^8 \text{ Å}}{1 \text{ cm}} = 3.28 \times 10^8 \text{ Å}$

5.2 a. $26.9 \ (\text{Å})^3 \times \dfrac{1 \text{ cm}^3}{10^{24} (\text{Å})^3} = 2.69 \times 10^{-23} \text{ cm}^3$

b. $2.69 \times 10^{-23} \text{ cm}^3 \times \dfrac{1 \text{ lit}}{1000 \text{ cm}^3} = 2.69 \times 10^{-26} \text{ lit}$

c. $2.69 \times 10^{-23} \text{ cm}^3 \times \dfrac{1 \text{ in}^3}{(2.54)^3 \text{ cm}^3} = 1.64 \times 10^{-24} \text{ in}^3$

5.3 a. $618 \text{ mm Hg} \times \dfrac{1 \text{ atm}}{760 \text{ mm Hg}} = 0.813 \text{ atm}$

b. $618 \text{ mm Hg} \times \dfrac{1.333 \times 10^3 \text{ dyne/cm}^2}{1 \text{ mm Hg}} = 8.24 \times 10^5 \text{ dyne/cm}^2$

5.4 $62.5 \dfrac{\text{lb}}{\text{ft}^3} \times \dfrac{454 \text{ g}}{1 \text{ lb}} \times \dfrac{1 \text{ ft}^3}{1728 \text{ in}^3} \times \dfrac{1 \text{ in}^3}{(2.54)^3 \text{ cm}^3} = 1.00 \text{ g/cm}^3$

5.5 $4.82 \times 10^4 \dfrac{\text{cm}}{\text{sec}} \times \dfrac{1 \text{ in}}{2.54 \text{ cm}} \times \dfrac{1 \text{ ft}}{12 \text{ in}} \times \dfrac{1 \text{ mile}}{5280 \text{ ft}} \times \dfrac{3600 \text{ sec}}{1 \text{ hr}}$

$= 1080 \text{ miles/hr}$

5.6 $\quad 16.0 \text{ g S} \times \dfrac{1.200 \text{ g metal}}{0.382 \text{ g S}} = 50.3 \text{ g}$

5.7 $\quad 1 \text{ mole OH}^- \times \dfrac{1.00 \text{ g acid}}{4.00 \times 10^{-3} \text{ mole OH}^-} = 250 \text{ g acid}$

5.8 $\quad 1 \text{ mole } e^- \times \dfrac{96500 \text{ coul.}}{1 \text{ mole } e^-} \times \dfrac{0.518 \text{ g}}{2.24 \times 10^3 \text{ coul}} = 22.3 \text{ g}$

5.9 $\quad$ a. $12.8 \text{ g Cu} \times \dfrac{1 \text{ GAW Cu}}{63.54 \text{ g Cu}} = 0.201 \text{ GAW Cu}$

$\quad$ b. $16.4 \text{ g Cu} \times \dfrac{1 \text{ GAW Cu}}{63.54 \text{ g Cu}} \times \dfrac{6.02 \times 10^{23} \text{ atoms}}{1 \text{ GAW Cu}}$

$$= 1.55 \times 10^{23} \text{ atoms}$$

$\quad$ c. $0.823 \text{ GAW Cu} \times \dfrac{63.54 \text{ g Cu}}{1 \text{ GAW Cu}} = 52.3 \text{ g Cu}$

5.10 $\quad$ a. $16.0 \text{ g H}_2\text{O} \times \dfrac{1 \text{ mole H}_2\text{O}}{18.0 \text{ g H}_2\text{O}} = 0.889 \text{ mole H}_2\text{O}$

$\quad$ b. $8.46 \times 10^9 \text{ molecules} \times \dfrac{1 \text{ mole}}{6.02 \times 10^{23} \text{ molecules}}$

$$\times \dfrac{18.0 \text{ g}}{1 \text{ mole}} = 2.53 \times 10^{-13} \text{ g}$$

$\quad$ c. $1.00 \text{ g H}_2\text{O} \times \dfrac{1 \text{ mole H}_2\text{O}}{18.0 \text{ g H}_2\text{O}}$

$$\times \dfrac{6.02 \times 10^{23} \text{ molecules}}{1 \text{ mole}} = 3.34 \times 10^{22} \text{ molecules}$$

5.11 $\quad$ a. $0.198 \text{ mole As} \times \dfrac{2 \text{ moles AsH}_3}{2 \text{ moles As}} = 0.198 \text{ mole AsH}_3$

$\quad$ b. $1.29 \times 10^{-4} \text{ mole AsH}_3 \times \dfrac{3 \text{ moles H}_2}{2 \text{ moles AsH}_3} = 1.93 \times 10^{-4} \text{ mole H}_2$

$\quad$ c. $1.48 \text{ mole H}_2 \times \dfrac{2 \text{ moles AsH}_3}{3 \text{ moles H}_2} \times \dfrac{77.9 \text{ g AsH}_3}{1 \text{ mole AsH}_3} = 76.9 \text{ g AsH}_3$

$\quad$ d. $16.0 \text{ g AsH}_3 \times \dfrac{1 \text{ mole AsH}_3}{77.9 \text{ g AsH}_3} \times \dfrac{2 \text{ moles As}}{2 \text{ moles AsH}_3} = 0.205 \text{ mole As}$

$\quad$ e. $0.619 \text{ g H}_2 \times \dfrac{1 \text{ mole H}_2}{2.02 \text{ g H}_2} \times \dfrac{2 \text{ moles As}}{3 \text{ moles H}_2} \times \dfrac{74.9 \text{ g As}}{1 \text{ mole As}} = 15.3 \text{ g As}$

$\quad$ f. $1.80 \text{ g AsH}_3 \times \dfrac{1 \text{ mole AsH}_3}{77.9 \text{ g AsH}_3} \times \dfrac{3 \text{ moles H}_2}{2 \text{ moles AsH}_3} \times \dfrac{2.02 \text{ g H}_2}{1 \text{ mole H}_2}$

$$= 0.0700 \text{ g H}_2$$

5.12   a. -107 kcal   b. -13.3 kcal   c. -36.2 kcal   d. -3.60×10^5 cal

  e. $1.24 \times 10^2 \text{g} \times \dfrac{-213 \text{ kcal}}{16.0 \text{ g}} \times \dfrac{1 \text{ BTU}}{0.252 \text{ kcal}} = -6.55 \times 10^3$ BTU

5.13   a. $4.08 - 2x$ b. $3x$ c. $28x$ 5.14   a. 4.52×10^8 kcal
                                                          b. 1.89×10^{19} erg

CHAPTER 6

Section 6.1

1. a. $x = 1.5 \times 10^{-4}$ c. $x = 1.33$ e. $x = 22/26$
    b. $x = 0.43$ d. $x = 4.40 \times 10^2$ f. $x = 12/7$
2. a. $V_2 = P_1 V_1 / P_2$ b. $T = PV/nR$ c. $\Delta H = 2.30 \, RT \, (B - \log P)$
    d. $\Delta E = RT \ln \dfrac{A}{k}$

Section 6.2

1. a. $x = -1/33, y = -59/33$ b. $x = 0.098, y = 0.102$
2. $x = 70/17, y = 7/34, z = 97/34$

Section 6.3

1. a. $x = \pm 1.4 \times 10^{-2}$ e. $x = 0.42 \text{ or } -0.72$
    b. $x = \pm 3.9 \times 10^{-5}$ f. $x = 6.7 \times 10^{-3} \text{ or } -6.8 \times 10^{-3}$
    c. $x = 3.4 \text{ or } 0.58$ g. $x = \pm 2.0 \times 10^{-4}$
    d. $x = 0.022 \text{ or } -0.023$ h. $x = 0.14 \text{ or } -0.15$
2. a. $x = 3.4 \times 10^{-2}$ b. $x = 4.2 \times 10^{-5}$

Section 6.4

1. a. $x = 1.0 \times 10^{-3}$ b. $x = 1.0 \times 10^{-2}$ c. $x = 6.4 \times 10^{-3}$
2. a. 0.05% b. 5% c. 1.5%;   2nd approx.: $0.000025\%, 0.25\%, 0.023\%$
3. a. 1st approx.: $x = 2.0 \times 10^{-2}$;   2nd approx.: $x = 2.0 \times 10^{-2}$
    b. $x = 0.98$
4. $x^2 + Kx - aK = 0$; $x = \dfrac{-K \pm \sqrt{K^2 + 4aK}}{2}$

  If $K \ll a, K^2 \ll 4aK$; $x = \dfrac{-K + 2\,(aK)^{1/2}}{2} = (aK)^{1/2} - K/2$

  This answer differs by $K/2$ from that obtained by making the approximation: $x = (aK)^{1/2}$. Percent error $= \dfrac{100 \,(K/2)}{(aK)^{1/2}} = 50 \,(K/a)^{1/2}$

Problems

6.1   a. (conc. Pb^{2+}) $(1.0 \times 10^{-2})^2 = 1.7 \times 10^{-5}$;   conc. $Pb^{2+} = 0.17$
      b. (1.0×10^{-1}) (conc. Cl^-)$^2 = 1.7 \times 10^{-5}$;   conc. $Cl^- =$
          1.3×10^{-2}
      c. (conc. Pb^{2+}) $(2 \text{ conc. } Pb^{2+})^2 = 1.7 \times 10^{-5}$;   conc. $Pb^{2+} =$
          1.6×10^{-2}

6.2 $\quad M = \dfrac{(1.86°C)\,(\text{no. of g. solute})}{(\text{f.p.l.})\,(\text{no. of kg water})} = \dfrac{1.86 \times 10.0}{0.372 \times 0.0900} = 55.6$

6.3 $\quad m = mX_2 + 55.6\,X_2; \qquad m = \dfrac{55.6\,X_2}{1 - X_2} = \dfrac{(55.6)\,(0.10)}{0.90} = 6.2$

6.4 $\quad \dfrac{x^2}{(0.50 - 2x)^2} = 0.20; \qquad \dfrac{x}{0.50 - 2x} = 0.45; \qquad x = 0.12$

6.5 $\quad \text{conc. } S^{2-} = \dfrac{1 \times 10^{-23}}{1 \times 10^{-18}} = 1 \times 10^{-5};$

$$\text{conc. } Ni^{2+} = \dfrac{1 \times 10^{-22}}{1 \times 10^{-5}} = 1 \times 10^{-17}$$

6.6 $\quad (\text{conc. } H^+) \times 5 = 1.80 \times 10^{-5}; \qquad \text{conc. } H^+ = 3.60 \times 10^{-6}$

6.7 a. $\dfrac{x^2}{1.0 - x} = 7.0 \times 10^{-4}; \qquad \dfrac{x^2}{1.0} = 7.0 \times 10^{-4}; \qquad x = 2.6 \times 10^{-2}$

b. $\% \text{ error} = 50\left(\dfrac{7.0 \times 10^{-4}}{1.0}\right)^{1/2} = 1.3\%$

c. $x^2 + 7.0 \times 10^{-4}x - 7.0 \times 10^{-4} = 0$

$$x = \dfrac{-7.0 \times 10^{-4} \pm \sqrt{49 \times 10^{-8} + 28 \times 10^{-4}}}{2}$$

$$\approx \dfrac{-7.0 \times 10^{-4} \pm \sqrt{28 \times 10^{-4}}}{2}$$

$$= \dfrac{-7.0 \times 10^{-4} + 5.3 \times 10^{-2}}{2} = 2.6 \times 10^{-2}$$

6.8 a. $\dfrac{x^2}{0.01 - x} = 1.80 \times 10^{-5};$

$x^2 = 1.80 \times 10^{-7} = 18.0 \times 10^{-8}; \qquad x = 4.24 \times 10^{-4}$

b. $\dfrac{x^2}{0.0100 - 0.0004} = 1.80 \times 10^{-5};$

$x^2 = 1.73 \times 10^{-7} = 17.3 \times 10^{-8}; \qquad x = 4.16 \times 10^{-4}$

c. Percent error in (a) $= 50\,(1.8 \times 10^{-3})^{1/2} = 2.1\%$
Fractional error in (b) $= 0.021 \times 0.021 = 0.00044 = 0.044\%$

6.9 $\quad \dfrac{x}{(1 - x)^2} = 1.8 \times 10^9;$

$x \approx 1,\, (1 - x)^2 = 1/(1.8 \times 10^9) = 5.6 \times 10^{-10}$

$1 - x = \text{conc. } H^+ = 2.4 \times 10^{-5}$

6.10 $\quad \dfrac{(2x)^2}{(2.0 - 2x)^2(1 - x)} = 1.0 \times 10^{-4};$

$4x^2 \approx 4 \times 10^{-4}; \qquad x = 1.0 \times 10^{-2}; \qquad \text{conc. } SO_3 = 2.0 \times 10^{-2}$

CHAPTER 7

Section 7.1

1. a. $\Delta S = (\Delta H - \Delta G)/T = 6.0\,\text{kcal}/300°\text{K} = +0.020\,\text{kcal}/°\text{K}$
    b. $\Delta G = 23.0\,\text{kcal} - 500°\text{K}\,(+0.020\,\text{kcal}/°\text{K}) = 13.0\,\text{kcal}$
    c. $0 = 23.0\,\text{kcal} - T\,(+0.020\,\text{kcal}/°\text{K}); \qquad T = 1150°\text{K}$

2. a. $\log \dfrac{K_2}{K_1} = 0; \qquad \dfrac{K_2}{K_1} = 1; \qquad K_2 = 1.0$

    b. $\log \dfrac{K_2}{K_1} = \dfrac{1.00 \times 10^4}{2.30 \times 1.99} \times \dfrac{(400 - 300)}{400 \times 300}$

    $= 1.82; \qquad K_2/K_1 = 66; \qquad K_2 = 66$

    c. $\log \dfrac{K_2}{K_1} = -1.82 = 0.18 - 2; \qquad K_2/K_1 = 0.015; \qquad K_2 = 0.015$

3. a. $R = \dfrac{PV}{nT} = \dfrac{(1\,\text{atm})\,(22.4\,\text{lit})}{(1\,\text{mole})\,(273°\text{K})} = 0.0821\,\dfrac{\text{lit atm}}{\text{mole}\,°\text{K}}$

    b. $0.0821\,\text{lit atm} \times \dfrac{24.2\,\text{cal}}{1\,\text{lit atm}} = 1.99\,\dfrac{\text{cal}}{\text{mole}\,°\text{K}}$

Section 7.2

1.

	a	b	c	d	e	f
	y	y	y	y	y	y
$x = 0$	0.0	∞	0.00	∞	12.0	-4.5
$x = 1$	6.0	12	2.10	1000	10.4	-3.0
$x = 2$	12.0	6.0	8.40	0.10	8.8	-1.5
$x = 3$	18.0	4.0	18.9	0.0047	7.2	-0.0
$x = 4$	24.0	3.0	33.6	0.0010	5.6	1.5
$x = 5$	30.0	2.4	52.5	0.00040	4.0	3.0

2. a. $s = 6.0, i = 0$ e. $s = -1.6, i = 12$ f. $s = 3/2, i = -9/2$

Section 7.3

1. a. y vs x^2 b. y vs $1/x$ c. $\log y$ vs x^2 d. $\log y$ vs $\log x$
    e. $y = (x - 1)^2$; y vs $(x - 1)^2$ f. $y = (x + 3/2)^2 - 5/4$;
    y vs $(x + 3/2)^2$
2. a. $a = 3/2, b = 0$ b. $a = 9, b = -27$ c. $a = -2, b = 8$
3. b. $\Sigma\ y = 5.0 + 7.8 + 10.8 + 14.0 + 16.8 + 19.8 = 74.2$
    $\Sigma\ x = 1.0 + 2.0 + 3.0 + 4.0 + 5.0 + 6.0 = 21.0$
    $\Sigma\ xy = 5.0 + 15.6 + 32.4 + 56.0 + 84.0 + 118.8 = 311.8$
    $\Sigma\ x^2 = 1.0 + 4.0 + 9.0 + 16.0 + 25.0 + 36.0 = 91.0$

    $a = \dfrac{(74.2)\,(21.0) - 6\,(311.8)}{(21.0)\,(21.0) - 6\,(91.0)} = 2.98;$

    $b = \dfrac{(311.8)\,(21.0) - 74.2\,(91.0)}{(21.0)\,(21.0) - 6\,(91.0)} = 1.95$

Problems

7.1   a. $\Delta S = -0.045 \text{ kcal/}°K$ b. -24.5 kcal c. $1040°K$

7.2   a. $\Delta E_{act} = \dfrac{2.303 \, RT_2 T_1}{(T_2 - T_1)} \log \dfrac{k_2}{k_1} = \dfrac{(2.30)(1.99)(658)(783)}{125} \log \dfrac{14.5}{0.276}$

$\qquad\qquad = 3.25 \times 10^4 \text{ cal}$

    b. $\log \dfrac{k_2}{0.276} = \dfrac{(3.25 \times 10^4)(42)}{(2.30)(1.99)(658)(700)} = 0.646; \qquad k_2 = 1.22$

7.3   Linear plot; intercept $= -47.0 \text{ kcal}$, slope $= 0.045 \text{ kcal/}°K$

7.4   b. You get a straight line with slope of 1.0×10^{-12}

$\qquad$ c. $(\text{conc. } CrO_4{}^{2-})(\text{conc. } Ag^+)^2 = 1.0 \times 10^{-12}$

7.5 $R = 0.0821 \text{ lit atm/mole }°K;$ a. $P = 0.00821 \, T$ b. PV indepen-

$\qquad$ dent of P; straight line parallel to x axis   c. $PV = 0.0821 \, T$

7.6   a. P vs $1/V$ b. m vs $1/u^2$ c. $\Delta G°$ vs $\log K_p$ d. $\log K_p$ vs $1/T$

$\qquad$ e. E vs $\log (\text{conc. } H^+)$ f. $\log (\text{conc. } A)$ vs T

7.7   a. $a = \dfrac{\log A_2 - \log A_1}{\Delta t} = \dfrac{-0.30}{4} = -0.075; \qquad \log 10.0 = 0 + b;$

$\qquad$ $b = 1.00$

$\qquad$ c.

$\log A$	t	$(\log A)\,t$	t^2
1.00	0	0.00	0
0.91	1	0.91	1
0.85	2	1.70	4
0.76	3	2.28	9
0.70	4	2.80	16
0.62	5	3.10	25
4.84	15	10.79	55

$\qquad$ $a = \dfrac{(4.84)(15) - 6(10.79)}{(15)(15) - 6(55)} = -0.075;$

$\qquad$ $b = \dfrac{(10.79)(15) - 4.84(55)}{(15)(15) - 6(55)} = 0.99$

7.8   b.

pH	$\log (\text{conc. } HA)$	$pH(\log \text{conc. } HA)$	$(\log \text{conc. } HA)^2$
2.4	0.0	0.0	0.0
2.9	−1.0	−2.9	1.0
3.4	−2.0	−6.8	4.0
3.8	−3.0	−11.4	9.0
12.5	−6.0	−21.1	14.0

$\qquad$ $a = \dfrac{(12.5)(-6.0) - 4(-21.1)}{(-6.0)(-6.0) - 4(14.0)} = -0.47; \qquad \log Ka = -4.84$

$\qquad$ $b = \dfrac{(-21.1)(-6.0) - (12.5)(14.0)}{(-6.0)(-6.0) - 4(14.0)} = +2.42 \qquad Ka = 1.5 \times 10^{-5}$

7.9   slope $\approx 1.67 \times 10^3; \Delta H_{vap} = (2.30)(1.99)(1.67 \times 10^3) = 7.64 \times 10^3 \text{ cal}$

7.10　Plots of c vs t or $1/c$ vs t are not linear. A plot of log c vs t gives a straight line with a slope of -0.10;　　$-k/2.3 = $ slope;　　$k = 0.23$

CHAPTER 8

Section 8.1

1. a. 0.4848　b. 0.9511　c. 0.9063　d. 1.150　e. 0.9272　f. -0.0872
2. $\sin^2 A + \cos^2 A = a^2/c^2 + b^2/c^2 = (a^2 + b^2)/c^2 = 1$
3. $a = (1.33\,\text{Å}) \sin 46° = 0.957\,\text{Å}$;　H—H dist. $= 1.91\,\text{Å}$
4. $a^2 = 4.00\,(\text{Å})^2 + 4.00\,(\text{Å})^2 - 2\,(2.00\,\text{Å})(2.00\,\text{Å})\cos 103° = 9.80\,(\text{Å})^2$
   $a = 3.13\,\text{Å}$;　　this problem could also be solved as in (3)
5. $a = 3.10\,\text{Å}$,　b $= 3.40\,\text{Å}$,　c $= 3.70\,\text{Å}$
   $(3.10\,\text{Å})^2 = (3.40\,\text{Å})^2 + (3.70\,\text{Å})^2 - 2\,(3.40\,\text{Å})(3.70\,\text{Å})\cos A$

$$\cos A = \frac{11.56 + 13.69 - 9.61}{25.16} = 0.622; \quad A = 52°$$

Similarly, $\cos B = 0.516$;　　$B = 59°$
$C = 180° - 52° - 59° = 69°$

Section 8.2

1. $3C$, 60°; $4C$, 90°; $5C$, 108°; $6C$, 120°. Would predict that cyclopentane would be most stable (closest to tetrahedral angle).

2. Area of face $= \dfrac{(0.120\,\text{cm})(0.120\,\text{cm})}{2} \sin 60° = 6.24 \times 10^{-3}\,\text{cm}^2$

Total area $= 2.50 \times 10^{-2}\,\text{cm}^2$

Section 8.3

1. The portion of each sphere that falls in the overlap region has a height of $0.2r$.

$$V\text{ segment} = \frac{\pi\,(0.04\,r^2)(2.8\,r)}{3} = \frac{0.112\,\pi r^3}{3}; \qquad V\text{ overlap} = \frac{0.224\,\pi r^3}{3}$$

Fraction $= \dfrac{0.224\,\pi r^3}{3} \Big/ \dfrac{8}{3}\pi r^3 = 0.028$

2. No. of atoms per cube $= 2$ (i.e., 1 in center, $\frac{1}{8}$ of each atom at corner)

V atoms $= 2 \times \dfrac{4}{3}\,\pi r^3 = \dfrac{8}{3}\,\pi r^3$;　length of body diagonal $= 4r = a\sqrt{3}$;

$a = 4r/\sqrt{3}$

V cube $= a^3 = \dfrac{64\,r^3}{3\sqrt{3}}$;　fraction of volume occupied $= \dfrac{\text{volume atoms}}{\text{volume cube}}$

$= \dfrac{\pi\sqrt{3}}{8} = 0.680$;　fraction of empty space $= 0.320$

3. Atoms touch along face diagonal, which has a length of $a\sqrt{2} = (2.90\,\text{Å})(1.414) = 4.10\,\text{Å}$. $4.10\,\text{Å} = 4r$;　$r = 1.03\,\text{Å}$
4. b—b: $4.00\,\text{Å}$, $2.83\,\text{Å}$;　　c—c: $3.00\,\text{Å}$;　　b—c: $2.50\,\text{Å}$

Problems

8.1 $d\,Cl\text{—}Cl/2 = (1.49\,\text{Å})\sin 58° = 1.264\,\text{Å}; \qquad d\,Cl\text{—}Cl = 2.53\,\text{Å}$

8.2 $(d\;Br\text{—}O)^2 = (2.14\,\text{Å})^2 + (1.15\,\text{Å})^2 - 2\,(2.14\,\text{Å})\,(1.15\,\text{Å})\cos 117°;$
$d\,Br\text{—}O = 2.85\,\text{Å}$

8.3 $\angle\,Cl\text{—}N\text{—}O = 180° - \dfrac{1}{2}\,(126°) = 117°$

$(d\,O\text{—}O)/2 = (1.24\,\text{Å})\sin 63° = 1.105\,\text{Å}; \qquad d\,O\text{—}O = 2.21\,\text{Å}$
$(d\;Cl\text{—}O)^2 = (1.79\,\text{Å})^2 + (1.24\,\text{Å})^2 - 2\,(1.79\,\text{Å})\,(1.24\,\text{Å})\cos 117°;$
$d\,Cl\text{—}O = 2.60\,\text{Å}$

8.4 $d\,H\text{—}O = 0.98\,\text{Å}\cos 76° + 1.46\,\text{Å} + 1.20\,\text{Å}\cos 64° = 2.22\,\text{Å}$

8.5 n = no. of atoms along one edge. $n\,(3.08 \times 10^{-8}\,\text{cm}) = 1\,\text{cm};$
$n = 3.25 \times 10^7$
Total no. of atoms = $n^2 = 1.06 \times 10^{15}$
Area occupied by argon atom = $\pi r^2 = 7.44 \times 10^{-16}\,\text{cm}^2$
Fraction covered = $(7.44 \times 10^{-16})\,(1.06 \times 10^{15})/1 = 0.789$

8.6 Consider a square with a corner at the center of four touching circles. Side of square = $3.08\,\text{Å}$. Length of diagonal = $4.34\,\text{Å}$. "Free" distance along diagonal = $4.34\,\text{Å} - 2\,(1.54\,\text{Å}) = 1.26\,\text{Å}$. $r = 1.26\,\text{Å}/2 = 0.63\,\text{Å}$.

8.7 $m = 65.4\,\text{g}/6.02 \times 10^{23} = 1.09 \times 10^{-22}\,\text{g}$

$V = \dfrac{4}{3}\,\pi\,(1.33 \times 10^{-8}\,\text{cm})^3 = 9.86 \times 10^{-24}\,\text{cc}$

$d = 109\,\text{g}/9.86\,\text{cc} = 11.1\,\text{g/cc}; \qquad f = 7.1/11.1 = 0.64\,(64\%)$

8.8 No. of atoms in cube = $8 \times \frac{1}{8} + 6 \times \frac{1}{2} = 4; \qquad V\,\text{atoms} = 16\,\pi r^3/3$
Face diagonal = $a\sqrt{2} = 4r; \qquad a = 4r/\sqrt{2} = 2r\sqrt{2}; \qquad V\,\text{cube} = 16\sqrt{2}\,r^3$

Fraction occupied = $\dfrac{\pi}{3\sqrt{2}} = 0.741; \qquad$ fraction empty space = 0.259

8.9 a. $1.50\,\text{Å}$ b. $4r = (3.00)\,(1.73); \qquad r = 1.30\,\text{Å}$
c. $4r = (3.00)\,(1.41); \qquad r = 1.06\,\text{Å}$

8.10 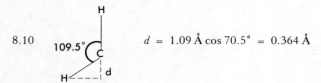 $d = 1.09\,\text{Å}\cos 70.5° = 0.364\,\text{Å}$

8.11 $4.66\,\text{Å}; \qquad 3.29\,\text{Å}$ 8.12 a. 1 b. 2 c. 5 d. 2 e. 15

CHAPTER 9

Section 9.2

1. $dy/dx = -8x = -8$
2. $dy/dx = 3x^2/8 = 27/8$
3. $dy/dx = e^x = 2.718$
4. $dy/dx = 1/x = 1/2$

5. $dy/dx = 1/2.30x = 0.362$
6. $dy/dx = \cos x = 0.866$
7. $dy/dx = -\sin x = -0.500$
8. $dy/dx = 3^x \ln 3 = 0.122$

Section 9.3

1. a. $y = u + v;$ $y + \Delta y = u + \Delta u + v + \Delta v;$ $\Delta y = \Delta u + \Delta v$

$$\frac{\Delta y}{\Delta x} = \frac{\Delta u}{\Delta x} + \frac{\Delta v}{\Delta x};$$ $$\text{Limit}_{\Delta x \to 0} \frac{\Delta y}{\Delta x} = \text{Limit}_{\Delta x \to 0} \frac{\Delta u}{\Delta x} + \text{Limit}_{\Delta x \to 0} \frac{\Delta v}{\Delta x};$$

$$\frac{dy}{dx} = \frac{du}{dx} + \frac{dv}{dx}$$

   b. $y = uv;$ $y + \Delta y = (u + \Delta u)(v + \Delta v);$
       $\Delta y = u\Delta v + v\Delta u + \Delta u \Delta v$

$$\frac{\Delta y}{\Delta x} = u\frac{\Delta v}{\Delta x} + v\frac{\Delta u}{\Delta x} + \frac{\Delta u \Delta v}{\Delta x};$$ $$\frac{dy}{dx} = u\frac{dv}{dx} + v\frac{du}{dx} + 0$$

2. $$\frac{d(u/v)}{dx} = \frac{ud(1/v)}{dx} + \frac{1}{v}\frac{du}{dx} = \frac{-u}{v^2}\frac{dv}{dx} + \frac{1}{v}\frac{du}{dx} = \frac{v\dfrac{du}{dx} - u\dfrac{dv}{dx}}{v^2}$$

3. a. $dy/dx = 5x^4 + 6x - 2;$ b. $dy/dx = x^2 \cos x + 2x \sin x + \sin x$
    c. $dy/dx = 3e^x \cos x - 3e^x \sin x;$ d. $dy/dx = 3(\ln x)^2/x$

    e. $$dy/dx = \frac{\cos x(\cos x) - \sin x(-\sin x)}{\cos^2 x} = \frac{\cos^2 x + \sin^2 x}{\cos^2 x} = \frac{1}{\cos^2 x}$$

    f. $$dy/dx = \frac{(e^x + 1)[2(e^x - 1)e^x] - (e^x - 1)^2 e^x}{(e^x + 1)^2} = \frac{e^{3x} + 2e^{2x} - 3e^x}{(e^x + 1)^2}$$

Section 9.4

1. $dy/dx = 2x + 6;$ $dy/dx = 0$ at $x = -3.$ At $x = -3.1, dy/dx = -0.2;$
    at $x = -2.9, dy/dx = +0.2.$ *Minimum*
2. $dy/dx = 2x - 6;$ $dy/dx = 0$ at $x = 3.$ At $x = 2.9, dy/dx = -0.2;$
    at $x = 3.1, dy/dx = +0.2.$ *Minimum*
3. $dy/dx = 3x^2 - 8x;$ $dy/dx = 0$ at $x = 0$ and $x = 8/3.$ At $x = -0.1,$
    $dy/dx = +0.83;$ at $x = +0.1, dy/dx = -0.77.$ *Maximum*. At $x = 7/3,$
    $dy/dx = -7/3;$ at $x = 9/3, dy/dx = 9/3.$ *Minimum*
4. $dy/dx = 3x^2 - 12x + 3.$ Using the quadratic formula, $dy/dx = 0$ at $x = 2 + \sqrt{3}$ or $2 - \sqrt{3},$ i.e., at $x = 3.73$ or $0.27.$ At $x = 3.6, dy/dx = -1.32;$ at $x = 3.8, dy/dx = +0.72.$ *Minimum*. At $x = 0.2, dy/dx = 0.72;$ at $x = 0.3, dy/dx = -0.33.$ *Maximum*
5. $dy/dx = \ln x + 1;$ $dy/dx = 0$ when $x = 0.367.$ At $x = 0.3, dy/dx = -0.20;$ at $x = 0.4, dy/dx = 0.08.$ *Minimum*
6. $dy/dx = \cos x;$ $dy/dx = 0$ at $x = 90°.$ Below $90°,$ $\cos x$ is positive; above $90°,$ $\cos x$ is negative. *Maximum*

Section 9.5

1. a. $2yx^2 dy + 2xy^2 dx = 0;$ $dy/dx = -y/x$

    b. $y^{1/2} dx + \dfrac{x}{2y^{1/2}} dy = 2ydy$

        $dy/dx = y^{1/2}/(2y - x/2y^{1/2})$
    c. $xdy/y + \ln y\, dx = 0;$ $dy/dx = -y(\ln y)/x$
    d. $y^2 e^x dx + e^x(2y) dy = 0;$ $dy/dx = -y/2$

2. $V = \frac{4}{3}\pi r^3$; $\quad \Delta V = 4\pi r^2 \Delta r = 4(3.14)(144.0)(0.01) \approx 18 \text{ cc}$

3. a. $\ln 1.010 = 0.010$; $\quad \log 1.010 = 0.010/2.3 = 0.0043$
   b. $\ln 0.990 = -0.010$;
      $\log 0.990 = -0.10/2.3 = -0.0043 = 0.9957 - 1$
   c. $\Delta \ln x = \Delta x/x = 0.030/2 = 0.015$;
      $\Delta \log x = 0.015/2.3 = 0.0065$; $\quad \log 2.03 = 0.3075$

4. a. 1.6×10^{-10}
   b. $\Delta \text{conc. Ag}^+ = 0.010$; $\quad \Delta K_{sp} = 0.010(1.6 \times 10^{-9}) = 1.6 \times 10^{-11}$
   c. $\Delta \text{conc. Cl}^- = 0.32 \times 10^{-9}$;
      $\Delta K_{sp} = (0.32 \times 10^{-9})(0.10) = 3.2 \times 10^{-11}$

5. $\rho = m/V$; $\quad \Delta \rho \approx -m\Delta V/V^2 = -\rho \Delta V/V$; $\quad \Delta \rho/\rho = -\Delta V/V$;
   $100 \Delta \rho/\rho = -100 \Delta V/V$

Problems

9.1 The rates should be approximately 1.0, 0.61, 0.37, and 0.22, respectively.

9.2 a. $\Delta°F = 1.8$; $\quad \Delta°F/\Delta°C = 1.8$
    b. $\Delta°F = 0.18$; $\quad \Delta°F/\Delta°C = 1.8$
    c. $d°F/d°C = 1.8$. Linear equation, constant slope.

9.3 $dV/dt = V_0(-4.5 \times 10^{-5} + 13.4 \times 10^{-6}t - 5.7 \times 10^{-8}t^2)$
    a. -4.5×10^{-5} $\quad$ c. $+25 \times 10^{-5}$
    b. $+0.8 \times 10^{-5}$ $\quad$ d. $+73 \times 10^{-5}$

9.4 a. $-(1.0 \times 10^{-14})/(\text{conc. OH}^-)^2$
    b. -1.0×10^{-14}, -1.0, -1.0×10^{14}

9.5 a. $-\Delta S°$
    b. $-\Delta S° = (\Delta G° - \Delta H°)/T$; $\quad -\Delta S° > 0$ if $\Delta G° > \Delta H°$

9.6 $dE/d(\text{conc. H}^+) = 0.059/\text{conc. H}^+$; $\quad$ a. 0.059 $\quad$ b. 5.9×10^5

9.7 $\dfrac{d\ln K_p}{dT} = -\dfrac{1}{R}\left[\dfrac{Td\Delta G°/dT - \Delta G°}{T^2}\right] = \dfrac{T\Delta S° + \Delta G°}{RT^2} = \dfrac{\Delta H°}{RT^2}$

9.8 $-dy/dt = k/(kt + c)^2 = ky^2$ $\quad$ 9.9 $-dA/dt = kA°e^{-kt} = kA$

9.10 $d(1/y^2) = d(2kt)$; $\quad -2dy/y^3 = 2kdt$; $\quad -dy/dt = ky^3$

9.11 $5.7 \times 10^{-8}t^2 - 13.4 \times 10^{-6}t + 4.5 \times 10^{-5} = 0$

$$t = \frac{13.4 \times 10^{-6} \pm \sqrt{180 \times 10^{-12} - 10 \times 10^{-12}}}{11.4 \times 10^{-8}} \approx 3.5°C \text{ or } 232°C$$

At $0°C$, dV/dt is negative; at $5°C$, it is positive. Minimum at $3.5°C$. The point at $232°C$ is fallacious; the equation does not apply at this high temperature.

9.12 $\Delta y \approx 2ax\,\Delta x = \dfrac{2y\,\Delta x}{x}$; $\quad \Delta y/y = 2\Delta x/x$; $\quad 100\Delta y/y = 200\Delta x/x$

9.13 a. $\Delta \log P \approx \dfrac{\Delta(-\Delta H \text{vap.})}{4.58(300)} = -\dfrac{100}{1374} = -0.073$;

     b. $\Delta \log P = \Delta P/P$; $\quad$ percent error $= 100\Delta P/P = -7.3\%$

CHAPTER 10

Section 10.2

1. a. $x^4/4 + C$ b. $e^x + C$ c. $5^x/\ln 5 + C$
   d. $3/(x + 3) + \ln (x + 3) + C$ (using formula 12)
2. a. $C = -1/4, \quad y = x^4/4 - 1/4; \qquad C = -e, \quad y = e^x - e;$
   $C = -5/\ln 5, \quad y = (5^x - 5)/\ln 5;$
   $C = -3/4 - \ln 4, \quad y = 3/(x + 3) - 3/4 + \ln (x + 3)/4$
   b. $C = 1 - 1/4 = 3/4, \quad y = (x^4 + 3)/4;$
   $C = 1 - e, \quad y = e^x + 1 - e;$
   $C = 1 - 5/\ln 5, \quad y = (5^x - 5)/\ln 5 + 1;$
   $C = 1 - 3/4 - \ln 4, \quad y = 3/(x + 3) + 1/4 + \ln (x + 3)/4$

Section 10.3

1. $3e^x + C$
2. Let $u = 6x; \qquad du = 6dx; \qquad dx = du/6;$

$$\int e^{6x} \, dx = \int e^u \, dx = \int \frac{e^u \, du}{6} = \frac{e^u}{6} + C = \frac{e^{6x}}{6} + C$$

3. $e^x + x \ln x - x + C$
4. Let $u = 1 - 2x; \qquad du = -2dx; \qquad dx = -du/2;$

$$\int \frac{dx}{(1 - 2x)^3} = \int \frac{dx}{u^3} = -\frac{1}{2} \int \frac{du}{u^3} = -\frac{1}{2} \, (u^{-2}/-2) + C$$

$$= \frac{1}{4 \, (1 - 2x)^2} + C$$

5. Let $u = 1 - x; \qquad du = -dx;$
   $$\int \ln (1 - x) \, dx = -\int \ln u \, du = -u \ln u + u + C$$
   $$= (1 - x)[1 - \ln (1 - x)] + C$$

6. $u = x + 2; \qquad du = dx;$
   $$\int (x + 2)^{1/2} \, dx = \int u^{1/2} \, du = \frac{2}{3} u^{3/2} + C = \frac{2}{3}(x + 2)^{3/2} + C$$

7. $u = e^x; \qquad du = e^x \, dx;$
   $$\int \frac{e^x}{e^x + 1} \, dx = \int \frac{du}{u + 1} = \ln (u + 1) + C = \ln (e^x + 1) + C$$

Section 10.4

1. a. $2^4/4 - 1/4 = 15/4; \qquad e^2 - e = 4.67;$
   $5^2/\ln 5 - 5/\ln 5 = 12.4; \qquad 3/5 + \ln 5 - 3/4 - \ln 4 = 0.073$
   b. $(-2)^4/4 - 3^4/4 - -65/4; \qquad e^{-2} - e^3 = 0.135 - 20.1 = -20.0;$
   $5^{-2}/\ln 5 - 5^3/\ln 5 \approx -5^3/\ln 5 = -77.7;$
   $3/1 + \ln 1 - 3/6 - \ln 6 = 0.71$

2. a. $\displaystyle\int_a^b F(x)\,dx = f(x = b) - f(x = a)$

$\displaystyle\int_b^a F(x)\,dx = f(x = a) - f(x = b) = -\int_a^b F(x)\,dx$

b. $\displaystyle\int_a^b F(x)\,dx + \int_b^c F(x)\,dx = f(x = b) - f(x = a)$

$+ f(x = c) - f(x = b)$

$= f(x = c) - f(x = a)$

$\displaystyle = \int_a^c F(x)\,dx$

Section 10.5

1. a. $\displaystyle\int_4^6 (3x - 6)\,dx = \left|\frac{3x^2}{2} - 6x\right|_4^6 = 18;$

b. $\displaystyle\int_0^2 (3x - 6)\,dx = \left|\frac{3x^2}{2} - 6x\right|_0^2 = -6$

c. $e - 1 = 1.72$

2. $\displaystyle\int_0^3 F(y)\,dy = \int_0^3 \frac{y}{2}\,dy = \left|\frac{y^2}{4}\right|_0^3 = 9/4$

Problems

10.1 a. $-dX/X = dt;$ $\quad -\ln X = kt + C;$ $\quad$ when $t = 0, X = X_0;$
$C = -\ln X_0;$ $\quad$ hence, $\ln X_0/X = kt$

b. $\ln 1.00/X = 0.100(10);$ $\quad \ln X = -1;$ $\quad X = 0.367$

c. $\ln \dfrac{X_0}{X_0/2} = kt_{1/2} = \ln 2;$ $\quad t_{1/2} = \dfrac{\ln 2}{k}$

10.2 a. $\displaystyle\int_{X_0}^{X} -dX/X^2 = \int_0^t k\,dt;$ $\quad 1/X - 1/X_0 = kt$

b. $1/0.500 - 1/1.00 = k(10.0);$ $\quad k = 0.100$

c. $2/X_0 - 1/X_0 = kt_{1/2};$ $\quad t_{1/2} = 1/kX_0$

10.3 a. $-dX/dt = k;$ $\quad$ c. $X_0 - X_0/2 = kt_{1/2};$ $\quad t_{1/2} = X_0/2k$
b. $X_0 - X = kt$

10.4 a. $\dfrac{-dX}{(k_1 + k_2)X - k_2 X_0} = dt$

Using formula 10, Table 10.1, with $a = k_1 + k_2, b = -k_2 X_0$:

$\dfrac{-1}{(k_1 + k_2)} \ln \dfrac{[(k_1 + k_2)X - k_2 X_0]}{[(k_1 + k_2)X_0 - k_2 X_0]} = t;$

$$\ln \frac{k_1 X_0}{(k_1 + k_2)X - k_2 X_0} = (k_1 + k_2)t$$

b. $\log \dfrac{0.100}{0.300X - 0.200} = \dfrac{0.300}{2.30} t;$

$\log(0.300X - 0.200) = -(1 + 0.130t)$

t	1	2	3	4
$\log(0.3X - 0.2)$	-1.130	-1.260	-1.390	-1.520
$(0.3X - 0.2)$	0.0741	0.0550	0.0407	0.0302
X	0.914	0.850	0.802	0.767

c. $X \rightarrow 2/3$

10.5 $-dX \dfrac{(a - X)}{X} = kdt; \qquad dX - a \dfrac{dX}{X} = kdt;$

$(X - X_0) - a \ln \dfrac{X}{X_0} = kt$

10.6 a. $\ln P = -\Delta H/RT + C$

b. $\ln \dfrac{P_2}{P_1} = \dfrac{-\Delta H}{R}\left[\dfrac{1}{T_2} - \dfrac{1}{T_1}\right] = \dfrac{\Delta H(T_2 - T_1)}{RT_2 T_1}$

10.7 a. $df = \dfrac{E_a}{RT^2} e^{-u} dT; \qquad du = \dfrac{-E_a}{RT^2} dT; \qquad df = -e^{-u} du;$

$f = e^{-u} + C; \qquad f = e^{-E_a/RT} + C$

As $T \rightarrow \infty, 1 = 1 + C; \qquad C = 0; \qquad f = e^{-E_a/RT}$

b. $\dfrac{d \ln f}{dT} = \dfrac{df}{f\,dt} = \dfrac{E_a}{RT^2}$

10.8 $\displaystyle\int_{300}^{500} (6.52 + 1.25 \times 10^{-3} T - 1.0 \times 10^{-9} T^2)\, dT$

$= 6.52(500 - 300) + \dfrac{1.25 \times 10^{-3}}{2}(500^2 - 300^2)$

$\qquad - \dfrac{1.0 \times 10^{-9}}{3}(500^3 - 300^3) = 1400$

10.9 $\displaystyle\int_{T_1}^{T_2} \dfrac{C_p}{T}\, dt = 6.52 \ln \dfrac{500}{300} + (1.25 \times 10^{-3})(500 - 300)$

$\qquad - \dfrac{1.0 \times 10^{-9}}{2}(500^2 - 300^2) = 3.58$

10.10 a. $W = P\Delta V = 9 \text{ lit atm}$ b. $W = 24.6 \ln 10 = 56.6 \text{ lit atm}$

10.11 Plot P vs V; take area from 1.0 lit to 10 lit. Approximately 70 lit atm

CHAPTER 11

Section 11.1

1.

Observation	Error	Deviation	Percent Error	Percent Deviation
328.1	0.9	0.5	+0.3	+0.2
327.6	0.4	0.0	+0.1	0.0
327.3	0.1	−0.3	+0.03	−0.1
327.4	0.2	−0.2	+0.06	−0.06

Note that changing to °K changes only the percent error and the percent deviation.

Trial	1	2	3	4	5	6	7	8	9
Error	+0.0105	+0.0102	+0.0104	+0.0103	+0.0105	+0.0104	+0.0106	+0.0119	+0.0099
Deviation	0.0000	−0.0003	−0.0001	−0.0002	0.0000	−0.0001	+0.0001	+0.0014	−0.0006

In this case, since the mean is obviously close to 2.3100, we might calculate it as:

$$2.3100 + \frac{(0.0002 - 0.0001 + 0.0001 + 0.0000 + 0.0002 + 0.0001 + 0.0003 + 0.0016 - 0.0004)}{9}$$

$$= 2.3100 + 0.0002 = 2.3102$$

Section 11.2

1. a. A and B correct; C too small. Hence, G.E.W. too high.
   b. C too large; G.E.W. too small.   c. C too large; G.E.W. too small.
2. a. $0.35/0.25 = 1.4$ b. $0.5/0.1 = 5$
3. a. 2050 Å.   0, −1, +1, +3, 0, +1, −2, +2, +4, −3, 0, +1, +2, −1, −2, −5, 0, −1, +1, 0.

   b.

y	1	0	1	2	3	5	4	2	1	1
x	−5	−4	−3	−2	−1	0	+1	+2	+3	+4

Note that the curve is not as smooth as it would be if n were much larger.

Section 11.3

1. a. Arith. mean $= 60.50$; $a = 0.18/4 = 0.045$;

   $$\sigma = \sqrt{\frac{(.00)^2 + (.09)^2 + (.03)^2 + (.04)^2 + (.02)^2}{4}}$$

   $$= 0.052$$

2. a. 112.30;   0.060   c. $0.674(0.060) = 0.040$; 112.30 ± 0.04
   b. ~68%
3. a. $M \pm 2.576 \sigma = 19.71 \pm 0.24$
   b. Dev. $= -2 \sigma$;   4.5%/2 ≈ 2%
4. a.

y	0.054	0.24	0.40	0.24	0.054	broad maximum
x	−2	−1	0	1	2	

   b.

y	0.0003	0.11	0.80	0.11	0.0003	narrow maximum
x	−2	−1	0	1	2	

Section 11.4

1. $M = 112.30 \pm 0.727 \dfrac{(0.060)}{\sqrt{6}} = 112.30 \pm 0.02;$

$M = 112.30 \pm 2.015 \dfrac{(0.060)}{\sqrt{6}} = 112.30 \pm 0.05;$

$M = 112.30 \pm 4.032 \dfrac{(0.060)}{\sqrt{6}} = 112.30 \pm 0.10$

2. $\sigma = \sqrt{\dfrac{(0.3)^2 + (0.3)^2}{1}} = \sqrt{0.18} = 0.42;$

$M = 16.9 \pm 63.66 \dfrac{(0.42)}{\sqrt{2}} = 16.9 \pm 19$

If he wants to be 99 per cent sure, he had better repeat the experiment!

3. Observed mean $= 201.2;$

$\sigma = \sqrt{\dfrac{(1.2)^2 + (2.4)^2 + (0.5)^2 + (0.4)^2}{3}} = 1.6$

$M = 201.2 \pm \dfrac{0.765\,(1.6)}{\sqrt{4}} = 201.2 \pm 0.6; \qquad \text{P.E.} = \dfrac{0.67\,(1.6)}{\sqrt{4}} = \pm 0.5$

Section 11.5

1. Trial mean (omitting 148.0) $= 152.3; \qquad a = 1.8;$
    $4.3 < 2.5\,(1.8) = 4.5$, *retain*; $\quad 4.3 < 4\,(1.8) = 7.2$, *retain*;
    $Q = 0.4/6.8 = 0.06 \ll 0.41$, *retain*.
2. 47.0, 49.4, 49.6, 50.1, 50.5, 55.2
    Test 55.2: $Q = 4.7/8.2 = 0.57 > 0.56$, *reject*.
    Test 47.0: $Q = 2.4/3.5 = 0.69 > 0.64$, *reject*.
    Test 50.5: $Q = 0.4/1.1 = 0.36 < 0.76$, *retain*.

Section 11.6

1. a. $d = m/V = 16.82\text{ g}/20.00\text{ ml} = 0.8410\text{ g/ml}$
        error in m: $E^2 = (.002)^2 + (.002)^2; \qquad E = 0.0028$
        error in d: $(E/0.8410)^2 = (0.0028/16.82)^2 + (0.01/20.00)^2$
        $= 28 \times 10^{-8}$
    $E = (5.3 \times 10^{-4})\,(0.8410) = \pm 0.0004$
2. $S = A + B; \qquad \Delta S = \Delta A + \Delta B; \qquad |\varepsilon_s| = |\varepsilon_a| + |\varepsilon_b|$
    $D = A - B; \qquad \Delta D = \Delta A - \Delta B; \qquad |\varepsilon_d| = |\varepsilon_a| + |\varepsilon_b|$

    $P = A \times B; \qquad \Delta P = A\Delta B + B\Delta A; \qquad \dfrac{\Delta P}{AB} = \dfrac{A\Delta B}{AB} + \dfrac{B\Delta A}{AB}$

    $\dfrac{\Delta P}{P} = \dfrac{\Delta B}{B} + \dfrac{\Delta A}{A}; \qquad \left|\dfrac{\varepsilon_p}{P}\right| = \left|\dfrac{\varepsilon_b}{B}\right| + \left|\dfrac{\varepsilon_a}{A}\right|$

$$Q = A/B; \qquad \Delta Q = \frac{B\Delta A - A\Delta B}{B^2}; \qquad \frac{\Delta Q}{A/B} = \frac{B\Delta A - A\Delta B}{AB}$$

$$\frac{\Delta Q}{Q} = \frac{\Delta A}{A} - \frac{\Delta B}{B}; \qquad \left|\frac{\mathcal{E}q}{Q}\right| = \left|\frac{\mathcal{E}a}{A}\right| + \left|\frac{\mathcal{E}b}{B}\right|$$

Problems

11.1   a. V too large; % Cl too large   b. A too large; % Cl too large
       c. M too large; % Cl too large   d.   % Cl too large

11.2   a.   36.1
       b. $-0.1, -0.3\%$; $+0.2, +0.6\%$; $-0.3, -0.8\%$; $+0.2, +0.6\%$
       c.   0.20   d.   0.24

11.3   a.   68%, 95%   b.   ±.03, ±.08, ±.16

11.4 $Q = \dfrac{1.21 - 1.12}{1.21 - 1.08} = 0.69 < 0.76;$ *retain*

11.5 $Q = \dfrac{56 - 39}{69 - 39} = 0.57 > 0.41;$ *reject*

11.6   Mean = 20.11, $\sigma = 0.10$;

$$M = 20.11 \pm \frac{1.83\,(0.10)}{\sqrt{10}} = 20.11 \pm 0.06$$

11.7 $M = \dfrac{1.545 \times 82.1 \times 298.5}{\dfrac{715}{760} \times 260} = 155$

$$\left(\frac{E_M}{155}\right)^2 = \left(\frac{0.001}{1.545}\right)^2 + \left(\frac{0.1}{298.5}\right)^2 + \left(\frac{1}{715}\right)^2 + \left(\frac{1}{260}\right)^2$$

$$= 4 \times 10^{-7} + 1 \times 10^{-7} + 2 \times 10^{-6} + 1.5 \times 10^{-5}$$

$$= 1.7 \times 10^{-5}; \quad E_M = \pm 0.6$$

INDEX

233

Find the Beat online!
Check us out at

www.shojobeat.com!

Find the Beat online!
www.shojobeat.com

BUSINESS REPLY MAIL

FIRST-CLASS MAIL PERMIT NO. 179 MT MORRIS IL

POSTAGE WILL BE PAID BY ADDRESSEE

NO POSTAGE
NECESSARY
IF MAILED
IN THE
UNITED STATES

SHOJO BEAT MAGAZINE
PO BOX 438
MOUNT MORRIS IL 61054-9964

WANTED
The Shojo Beat Manga Edition

STORY AND ART BY
Matsuri Hino

English Adaptation/NANCY THISTLETHWAITE	Editor in Chief, Books/ALVIN LU
Translation/TOMO KIMURA	Editor in Chief, Magazines/MARC WEIDENBAUM
Touch-up Art & Lettering/SABRINA HEEP	VP of Publishing Licensing/RIKA INOUYE
Design/COURTNEY UTT	VP of Sales/GONZALO FERREYRA
Logo Design/AARON CRUSE	Sr. VP of Marketing/LIZA COPPOLA
Editor/NANCY THISTLETHWAITE	Publisher/HYOE NARITA

Printed in Canada

PUBLISHED BY VIZ MEDIA, LLC
P.O. BOX 77010
SAN FRANCISCO, CA 94107

SHOJO BEAT MANGA EDITION
10 9 8 7 6 5 4 3 2 1
FIRST PRINTING, SEPTEMBER 2008

store.viz.com

MATSURI HINO BURST ONTO THE MANGA SCENE WITH HER TITLE *Kono Yume ga Sametara* (WHEN THIS DREAM IS OVER), WHICH WAS PUBLISHED IN *LaLa DX* MAGAZINE. HINO WAS A MANGA ARTIST A MERE NINE MONTHS AFTER SHE DECIDED TO BECOME ONE.

WITH THE SUCCESS OF HER POPULAR SERIES *Captive Hearts* AND *MeruPuri*, HINO HAS ESTABLISHED HERSELF AS A MAJOR PLAYER IN THE WORLD OF SHOJO MANGA. *Vampire Knight* IS CURRENTLY SERIALIZED IN *LaLa* AND *Shojo Beat* MAGAZINES.

HINO ENJOYS CREATIVE ACTIVITIES AND HAS COMMENTED THAT SHE WOULD HAVE BEEN EITHER AN ARCHITECT OR AN APPRENTICE TO TRADITIONAL JAPANESE CRAFT MASTERS IF SHE HAD NOT BECOME A MANGA ARTIST.

⚓ Afterword ⚓

For some reason, I still have lots of ideas for Wanted. For example, I'd like to draw a story where a female pirate appears...my thoughts are smoldering. I just started Vampire Knight in LaLa, so I must concentrate on that, though...

A sequel...whether I can do one is undecided, and I don't know how things will turn out (I won't be able to unless I can come up with interesting stories ♪), but I'd be happy if I can see all of you in Wanted again! And finally, thank you for reading this manga. Please read Vampire Knight too! ♡

Matsuri Hino

My heartfelt thanks to my editor, who's been taking good care of me; to I-san and to my mother, who helps me out every month; and to S.U.-sama and A.I.-sama, who I ended up asking for help in a rush.

SPRING
CHERRY
BLOSSOMS

I CAME BACK TO GET SOMETHING.

PLEASE
TAKE
MY HAND
AGAIN...

THROW THEM OUT ON THE STREET.

SURE THING.

fwup fwup fwup

SWIp

klak

THANK YOU FOR RESCUING ME.

WILL YOU ESCORT ME?

...BUT I'M LOST.

I WAS ON MY WAY HOME...

WE'RE BORROW-ING THIS.

IF YOU CAN, I'LL BREAK THE ENGAGEMENT.

...THOUGH IT'S A STRANGE BET.

I ACCEPTED...

TAKAO...

HE DIDN'T COME BACK TO SAKURADAYA LAST NIGHT.

YOU COME WITH US.

HEY!

GRAB

TAKAO BEAT THEM UP YESTERDAY.

...HE WAS CARRYING ON HIS BACK LAST NIGHT.

AH, THERE'S THE WOMAN...

OTHER- WISE WE'LL BREAK THIS WOMAN'S ARM!

WE WANT TO TALK TO THE WIDOW!

!!

...AND THE STAFF UNDER MY MOTHER WANTED ME TO TAKE OVER...

...THE STAFF UNDER THE GENERAL MANAGER WANTED TAKAO TO TAKE OVER...

BUT...

I'LL HELP BIG BROTHER A LOT!

I wish I could've seen that.

Adorable, right?

...SO THE STORE STAFF BECAME DIVIDED.

HE WAS FULL OF ENERGY.

AND IT'S NOT BECAUSE DAD'S WILL SAID SO! I'M SO HAPPY I HAVE A BROTHER NOW!

EVER SINCE THEN HE'S BEEN ACTING THIS WAY.

...MIGHT GO OUT OF BUSINESS IF THE BICKERING CONTINUED.

TAKAO'S SHARP. HE REALIZED THE STORE...

I NEVER WANTED THE STORE...

YES.

I LIKE BOLD WOMEN.

YOU MUST KNOW WHY I RAN AWAY FROM HOME.

I MET TAKAO IN MY TOWN.

HE'S GOING TO KEEP TELLING ME LIES FOR HIS BROTHER'S SAKE.

I'M HAPPY THAT YOU CARE ABOUT TAKAO.

HE'S NOT COMPLETELY RECKLESS.

IF HE SAID HE'S FINE, HIS WOUND ISN'T SERIOUS.

BUT THE ONE WHO TOOK MY HAND FIRMLY ...

THE ONE IN MY THOUGHTS ...

...IS YOU.

He's adorable, isn't he?

heh heh

He's my dear brother.

YOU UNDERSTAND HIM WELL...

I LOVED MY HALF-BROTHER FROM THE FIRST TIME WE MET.

I'VE BEEN TOLD...

...THAT WE ARE BETROTHED.

I'M TAKAO'S OLDER BROTHER, KYOSUKE SAKU-RADA. I'M MASTER APPRENTICE OF SAKURADAYA, THIS WHOLE-SALE STORE.

I UNDER-STAND.

...AND MY FUTURE HUSBAND. BUT IT'S NOT MY CHOICE.

TAKAO'S...

...OLDER BROTHER...

FWAP

THE STORE STAFF EXPECTED HIM TO TAKE OVER SAKURADAYA BECAUSE I WASN'T IN GOOD HEALTH...

MY LITTLE BROTHER IS ACTUALLY A SMART AND KIND BOY.

WAIT!

skrtch skrtch

...BUT NOW THAT THE PUBLIC CONSIDERS HIM A HOODLUM...

PLEASE LET A DOCTOR TREAT THAT WOUND!

THERE YOU GO AGAIN.

BIG BROTHER IS SAD.

...SO PEOPLE MIGHT GOSSIP ABOUT HER...

SHE WAS SEEN ON THE STREETS WITH ME...

TAKAO ...?

SORRY, BROTHER.

TAKAO, YOU'RE HURT. THAT RARELY HAPPENS. YOU WON, RIGHT?

I APOLO- GIZE FOR MY YOUNGER BROTHER'S RUDENESS, SHO KAMURA.

LET ME INTRO- DUCE MYSELF.

Stop teasing her, Takao.

VUMP

IT'S MY FAULT! HE TOLD ME TO GO HOME, BUT I DIDN'T.

NO!

I DIDN'T DO IT FOR FUN!

Heh

YOU REALLY ENJOYED RUNNING AWAY FROM HOME.

WITH HER REPUTATION RUINED, NO ONE WILL WANT TO MARRY HER ...

PSH PSH

STARE

LOOK! THIS TIME HE'S WITH A YOUNG LADY!

PSH PSH

There, there.

SHALL I COMFORT YOU?

WE MUST SEND A MESSENGER TO THE KAMURAS, LETTING THEM KNOW THAT SHE'S STAYING WITH US TONIGHT!

SHE'S KYOSUKE'S FIANCÉE!

SAKURA

THIS IS WHY A MISTRESS'S CHILD...!

I DON'T NEED IT.

RUB

THAT OLD WOMAN ACTS LIKE A WIFE...

I HOPE YOU HAVEN'T DONE ANYTHING TO HER!

SHE RAN AWAY FROM HOME ?!

THIS GIRL IS SHO KAMURA!

WHY...

RIP. RIP.

SHCK

DON'T MOVE!

BUT THIS IS THE FIRST TIME I'M TREATING A WOUND...

AFTER THE BLEEDING STOPS, CALL A DOCTOR...

FAINT

Sigh

SORRY, CHIYO.

I'M GOING HOME TONIGHT.

Ah!

Psst Psst

He got into a fight again...

Seventh

"Spring Cherry Blossoms: A Small Incident at Sakuradaya, Meiji Era"

I drew this story in between Captive Hearts and Meru-Puri. My drawing is different. It's close to the style I used when I began MeruPuri. In Meru-Puri, I made the expressions sweet, sweeeeeet, and gradually made them close to my ideal.

This story was born from my ambitions of drawing kimonos and an idiot!! I really love the combination of Takao's gaudy haori half-coat and the ten-gallon hat.

This is just a one-shot, but I like Sho-chan too. I also like Takao's older brother. (I really love writing about brothers.) (smile)

When I reread this story, I realized once again that "Spring Cherry Blossoms" is the base material for Wanted and MeruPuri.

KLAK
KLAK

huff

huff

MEIJI ERA (1868–1912): CULTURE BLOOMED BACK THEN.

Peek

THAT'S A NEW AND VERY DIFFICULT BOOK.

"COMMERCE AFTER THE RESTORATION AND THE CURRENT STATE OF FOREIGN COUNTRIES" ...?

Peek

HE'S DRESSED SO FLAM-BOYANTLY, AND HE'S READING A BOOK LIKE THAT?

What an odd hat...

Skulls and *MeruPuri*'s
Aram and Jeile were
born because I was
able to create Takao
in "Spring Cherry
Blossoms."

Aram
&
Jeile

SPRING CHERRY BLOSSOMS

A Small Incident at Sakuradaya, Meiji Era

GLARE

I'LL KILL YOU FOOLS IF YOU DON'T SHUT UP.

I DON'T REMEMBER ANYTHING LIKE THAT.

GYAAAH—
Captain has gone berserk!

He's angry!
Yeeeek

I'LL PRETEND I DIDN'T HEAR ANY-THING.

TROMP TROMP TROMP

DOC, YOU...

THE FEVER MADE HIM BLURT OUT HIS TRUE FEELINGS.

At death's precipice.

ho ho ho

THOUGH I'D BE HAPPY IF HE MEANT IT.

The Legend of the Devil's Musical Score/End

SKULLS
...

...THIS PIECE IS TRULY GLORIOUS IN ITS BEAUTY, SO...

THE NAVY SHOULD BE HELPING PEOPLE IN NEED!

...HE'S STILL DISTURBING THE PEACE AT SEA.

BUT...

I'VE HEARD STORIES ABOUT THE TWO CAPTAIN SKULLS HELPING COMMONERS.

DO YOU UNDER-STAND?

THERE IS NO JUSTIFI-CATION FOR PIRACY.

THAT MAN...

...IS HAVING FUN COMMITTING CRIMES WHILE PRETENDING TO BE ROBIN HOOD.

SHNNK

...

OH

TMP

AN ARROW FROM THE TRAP PIERCED YOU...

I'M ALL RIGHT.

YOU'RE BLEEDING!

SKULLS!

SKULLS... THIS LOOKS LIKE A POISONED ARROW.

KRYSH

KRAK

KRIK KRIK KRIK

THOOM

KRAAH!

TUNK TUNK

DAMN.

SOME FOOL SHOT A CANNON.

YOU...

I COULDN'T HELP IT. I HAD BUSINESS HERE...

...THE SAME AS YOU.

THE NAVY IS PATROLLING DESERTED ISLANDS NOW?

HE'S THE ONE WHO CAPTURED ME THAT TIME.

THIS MAN!

ARTO, YOU'LL OWE ME FOR THIS.

sigh

I HOPE YOU'LL DO AS I SAY?

THIS WAY, PLEASE.

BUT THIS IS QUITE CONVENIENT.

...

LIAR.

shk shk

Stubborn

I am not!

YOU'RE SCARED OF FINDING THE SCORE NOW, AREN'T YOU?

It's the Devil's curse!

NOW I KNOW HOW STUBBORN A SONGSTRESS CAN BE... JUST DON'T STUMBLE IN YOUR HASTE!

I'M NOT SCARED! I'LL SING THAT LEGENDARY SONG!

GLOM

Be careful.

grin

Oh

THERE'S NO SUCH THING.

Sixth

Skulls-sama's Merry Comrades

Death shall seize your skull

WELL?

WHERE'S YOUR COM-MANDER?

...

He said he'd search the caves until our ship returned!

RIGHT. I'LL STRING YOU UP AGAIN AND GET THE CATER-PILLARS—

The tattoo.

It's heartwarming to think that Doc and Luce have the same tattoo.

The meaning is as Reid explained it.

If I must compliment his explanation, it's a mark of readiness to "always be facing death."

KYAAH

ON THE SEA...

...WE PIRATES ARE MORE SAVAGE THAN THE DEVIL.

grin

CAP-TAIN!

WHAT SHOULD WE DO IF WE'RE CURSED BY THE DEVIL?!

GRIN

AYE, CAP-TAIN...

THEY ALL AGREE.

They think it's fun.

PIRATES REALLY ARE UP TO NO GOOD!

TO GO UP AGAINST THE DEVIL...

ALL RIGHT! THIS MUST BE THE ISLAND!

NAVY
VERMIN.

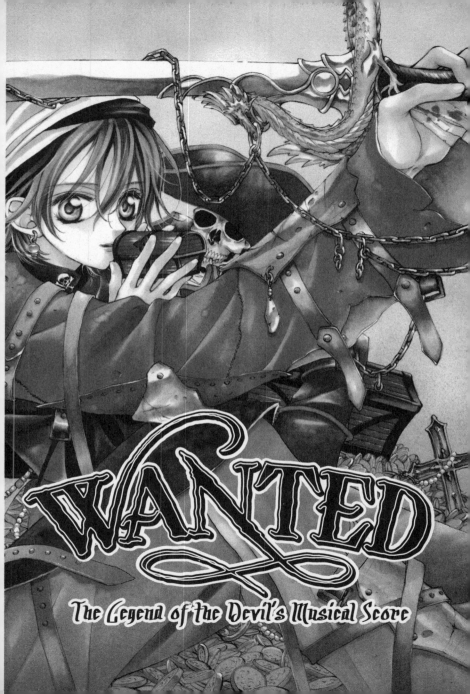

When Armeria slept next to Skulls-sama, his heart whispered:

Silly girl, I'll force myself on you!

No, don't do it. Get ahold of yourself!

YOU'RE SHY, RIGHT?

heh

WE CAN ALWAYS BE TOGETHER, RIGHT, SKULLS?

AH...

YOU LAUGHED.

The Enemies of Pirates/End

THUD

I SHOULD ONLY WORRY ABOUT NOT BEING ABLE TO SEE HIM AGAIN, NOT WHETHER HE'D ABANDON ME!

OH...

AH! THE COMMANDER WILL THINK I DID SOMETHING TO HER.

OF ALL THE ROTTEN LUCK!

SSSsp

I TRAVELED ALONE TO FIND LUCE. I CAN DO THIS ON MY OWN!

ARE YOU ALL RIGHT, MISS? ARE YOU HURT?

KREE

CHAK

TMP

GEH.

AS A FAVOR TO THE LASS, I'LL GO.

IF YOU'RE A SAILOR, YOU SHOULD BE LEAVING FROM THE FRONT DOOR.

JOLT

AND YOU, YOUNG LADY.

DASH

WHAT WOULD COMPEL YOU TO YELL THAT SKULLS ISN'T HERE?

FOL-LOW HIM!

YES SIR!

CAPTAIN!

I WONDER HOW MANY ARE STATIONED THERE.

THAT'S THE NAVY FORT...

MOST OF THE SHIPS ARE OUT TO SEA ...

THAT'S THE PORT OF PERINA.

IT'S FAMOUS FOR BEING A NAVY STRONG-HOLD.

WE CAN TAKE A BOAT TO SHORE.

Right under their noses.

ARTO'S GONE! SHE TOOK A BOAT!

THEY'RE CRACKING DOWN ON PIRATES.

THE WAY THEY KEPT COMING AFTER US THE OTHER DAY, THEY'RE TAKING THE JOB SERIOUSLY.

Third

Skulls-sama's Merry Comrades

...

Ahoy!

FUMP

THERE BE A GOOD PLACE TO HIDE THE SHIP FAR FROM PORT.

Doc

He's the original Pirate Skulls and is now the onboard doctor.

"When you be troubled, ask Doc." The crew depends on him.

I'd like to draw the chapter where he lets Luce take over as captain...

I HAVEN'T BEEN SLEEPING WELL LATELY...

DOC.

YOU SHOULDN'T HAVE ALLOWED ME ONBOARD THEN!

NO MATTER WHAT I TRY TO DO, YOU'RE AGAINST IT!

I'M GOING TO BED!

A BEAUTIFUL MEMORY, MILOR—

YOU DON'T KNOW HOW YOU SHOULD TREAT HER, DO YOU?

YOU STILL SEE HER AS THE LITTLE SINGING PRINCESS FROM YOUR CHILDHOOD.

ARTO THREW DISHES AT THE CAPTAIN'S HEAD.

LASSES BE SCARY WHEN MAD.

SHE'S JUST THROWING A TANTRUM.

...I HAD GLORIFIED LUCE IN MY MEMORY...

HEY, ARTO.

ABOUT WHAT I SAID EARLIER.

YOU DIDN'T HEED MY WORDS.

tug

CAP'N STEPPED FORWARD TO PROTECT YOU, ARTO.

A CAPTAIN'S ORDER IS ABSO-LUTE.

YOU BEST REMEMBER THAT IF YOU WANT TO STAY ON THIS SHIP.

THIS PIRATE LOOKED SO EVIL THAT I COULDN'T SEE THAT HE WAS LUCE.

YOU'RE IMAGINING THINGS, YOU FOOL!

THEY GOT AWAY.

DOES HE KNOW EVERYTHING ABOUT THE SEA?

THAT PIRATE SKULLS...

HEY, ARTO! LEAVE THAT TO THE MEN!

THAT BE PRETTY HEAVY, LASS. CAREFUL, NOW.

I'M 15, BUT I DISGUISED MYSELF AS A BOY AND BOARDED THIS PIRATE SHIP.

I'M FINE!

MY REAL NAME IS ARMERIA.

THIS? I'M JUST TAKING IT OVER THERE.

When Skulls-sama
cried out Armeria's name,
his heart whispered:
SILENCE

I can't look her in the eye.
I just revealed who I am.
I'm an idiot!

Captain Skulls/End

PIRATES ARE DAMNED LIARS...

A LONG TIME AGO, HE WATCHED HIS UNCLE, GOVERNOR LANCEMAN...

...STEAL FROM THE PEASANTS, JUST LIKE A PIRATE.

HE REGRETTED NOT DOING SOMETHING ABOUT IT.

OH?

THEN YOU WERE...

Ho ho!

HE TOOK THE NAME "SKULLS" AWAY FROM ME...

SOON HE STARTED SAYING THAT HE'D BECOME A STRANGE PIRATE LIKE ME, A PIRATE WHO'D HELP THE WEAK...

AFTER THAT ATTACK, WE ENDED UP HAVING TO RETURN TO SEA WITH HIM ONBOARD.

SKULLS!

hukk
hukk
hukk

WHO'S THERE?

I'M ON DUTY...

BUT WHEN HE FEELS LIKE IT, HE HELPS TOWNS THAT HAVE BEEN PILLAGED BY PIRATES. HE ALSO HELPS PEASANTS SUFFERING UNDER UNJUST LANDOWNERS.

THE WOMEN DRESSED ME UP IN ALL THIS!

F*wip*
F*wip*

Uh!

SHUT UP!

heh

YOU LOOK RIDICU-LOUS.

NO.

DON'T BELIEVE THOSE STORIES.

They're rubbish.

I WAS TOLD YOU ACTUALLY HELP PEOPLE!

I DON'T BELIEVE IT!

...IF YOU INTEND TO LOOK FOR LUCE IN THIS TOWN...

BY THE WAY, ARTO...

DID HE SMILE?

SKULLS?

klap klapklap

klap klap klap

That cheered me heart, Arto

YES, BUT HE'S A STRANGE PIRATE.

HE DOESN'T ROB PEOPLE LIKE US.

THINGS ARE LIVELY NOW!

MUCH LIKE WHEN SKULLS FIRST CAME TO REBUILD THIS TOWN.

Huh?!

MADAM, HE'S A PIRATE!

!

YOU WERE WONDERFUL! DRESS UP AND SING AGAIN.

YOU'RE CHARMING! I'LL GIVE YOU ONE OF MY GOWNS!

SHOW THEM, ARTO! YOUR VOICE BE LIKE A SIREN'S!

STAND OVER HERE AND LET ME HEAR YOU SING.

AYE, I WANTED TO HEAR THE LASS SING TOO!

I DON'T WANT TO SING IN A PLACE LIKE THIS...

MRR

HE'S TOO OCCUPIED TO LISTEN TO MY SONG.

Second

Skulls-sama's Merry Comrades

He's the second-in-command.

He knows Skulls well—long before Skulls became captain.

He's dependable, easy to talk to, and he's fun to be around.

He's from a kingdom in the Southern Islands. He might have been a monk previously...?

WHAT A LECHER!

He's surrounded by women...

Peek

HEY, ARTO...

...GET OVER HERE.

blush

!

STOP YANKING ME AROUND!

tug

THNK

PORT DE FLEUR

WE'RE GETTING OFF.

I'VE NEVER BEEN TO THIS TOWN BEFORE.

MAYBE I CAN FIND OUT SOMETHING ABOUT LUCE...

WHY WOULD THE TOWNSFOLK WELCOME PIRATES?

HUH?!

WE MISSED YOU!

WELCOME HOME, CAPTAIN SKULLS!

!

THUD

HE WILLFULLY SWINDLED THE COMMON PEOPLE AND DEALT IN DIRTY BUSINESS, ALL UNDER THE ORDER OF THE MARQUIS.

HOW CAN YOU KILL PEOPLE SO EASILY?

HE WAS AN EVIL MAN.

IS THAT WHY YOU KILLED HIM?

HE WAS A MINION OF THE MARQUIS.

HE... SAVED ME AGAIN.

...THE MARQUIS OF GLENGER'S...

...GAUDY TREASURE SHIP!

I KNEW OUR CAPTAIN WOULD HEAR IT.

SHE'S OVER HERE...

ARE THEY FIGHTING?

WHAT'S GOING ON?!

YOU SAID YOU'VE NEVER SEEN IT, SO...

THIS IS ARMERIA, THE FLOWER THAT IS YOUR NAMESAKE.

LUCE PICKED THIS ARMERIA FOR ME...

THE PETALS OF THAT FLOWER HAVE BEEN IN MY LOCKET THESE PAST EIGHT YEARS...

IF I CONTINUE SEARCHING...

...I'LL BE ABLE TO SEE YOU AGAIN, RIGHT?

First

「WANTED」

A pirate manga... I've always wanted to draw one, even before I made my debut. When I got the go-ahead for the first chapter, my satisfaction was immense. (smile) And, fortunately, I was able to do a complete volume.

Arto is a girl, and Skulls is actually --. These details are pretty obvious from the beginning, but they are not the important part of this story, so it's all right. (smile) (I was defiant from the start when I was working on my storyboards...)

By the way, Skulls is so arrogant that I call him Skulls-sama at times. Arto (Armeria) is the female version of MeruPuri's Aram... (She's supposed to be like him, but she's such a commoner, she just doesn't look it...♪)

I did study the various historical backgrounds (more or less♪), but I've ignored them to make things convenient. Hmph...It's fiction anyway. ♪

DON'T THINK I'M PROTECTING YOUR HONOR, ARTO.

SKULLS...

I CAN'T DO IT.

IT'S JUST YOU'RE AS SCRAWNY AS A LAD.

YOU'RE HORRIBLE! YOU'RE NOTHING COMPARED TO LUCE!

HMPH.

SCRAWNY AS A LAD ?!

I Um... WANT TO ASK YOU ABOUT LUCE...

DOES HE KNOW WHAT HAP-PENED?

THANK YOU...

YOU CAN REMOVE THE BANDAGES IN TWO OR THREE DAYS.

YOU WERE LUCKY THE BULLET ONLY GRAZED YOU.

WELL... I KNEW SKULLS BEFORE HE BECAME CAPTAIN.

HE'S A PIRATE TOO.

smile

Tup

DID YOU THANK HIM?

...

chak

AND HE SAVED YOU.

HE'S BEEN A GOOD CAPTAIN TO US.

WHY WOULD I THANK HIM?

THE CAPTAIN SAID IT'S FOR THE SAFETY OF THE CREW...

?

WHAT ...

...IS THIS ROPE FOR?

HE'S FINE!

HE'S STILL BREATH-ING.

PULL HIM UP!

SPUSH

EHH?!

WILL HE BE ALL RIGHT?

UH...

THIS LAD BE LIGHT AS A FEATHER!

He barely eats.

FORGIVE ME, ARTO!

YOU APOLO-GIZE LATER!

I'M NOT SHARK FEED YET...

HMPH! THAT SHOULD SHOW HIM...

SLUMP

SO HE IS CAPTAIN SKULLS.

HE'S A DESPICABLE PIRATE, JUST AS I'D IMAGINED.

I CAN'T BELIEVE I THOUGHT SKULLS WAS LUCE—THE ONLY THING THEY HAVE IN COMMON IS THE SAME HAIR COLOR.

I'M SORRY, LUCE.

I HAVE TO KEEP GOING UNTIL I FIND LUCE...

fump

LUCE ...

HOW SHOULD I APPROACH THE TOPIC OF LUCE TO SKULLS?

EVEN NOW, EIGHT YEARS LATER...

...I STILL REMEMBER YOUR KINDNESS.

GOVERNOR LANCEMAN'S ESTATE OVERLOOKING THE MEDITERRANEAN

LATE 17TH CENTURY

YOUR EXCELLENCY, ARMERIA'S THROAT IS ALREADY—

SING THAT SONG ONCE MORE.

GIRL.

he! he!

I'M ALL RIGHT, MISTRESS.

SINGING IS MY TRADE.

UNCLE!

STOP TORTURING HER! ENOUGH IS ENOUGH!

...MY 30TH SONG.

I LOST MY FAMILY WHEN I WAS VERY YOUNG. I JOINED A MUSICAL TROUPE THAT PERFORMED ON THE ESTATES OF ARISTOCRATS IN VARIOUS AREAS.

WANTED

Captain Skulls

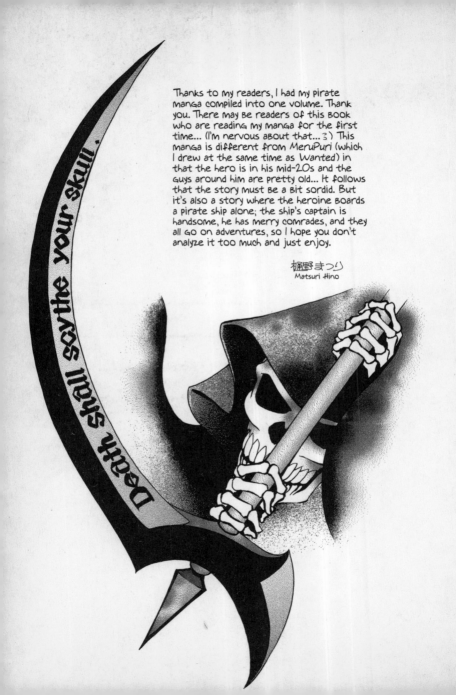

Thanks to my readers, I had my pirate manga compiled into one volume. Thank you. There may be readers of this book who are reading my manga for the first time... (I'm nervous about that... ♪) This manga is different from MeruPuri (which I drew at the same time as Wanted) in that the hero is in his mid-20s and the guys around him are pretty old... It follows that the story must be a bit sordid. But it's also a story where the heroine boards a pirate ship alone; the ship's captain is handsome, he has merry comrades, and they all go on adventures, so I hope you don't analyze it too much and just enjoy.

樋野まつり
Matsuri Hino

Death shall scythe your skull.

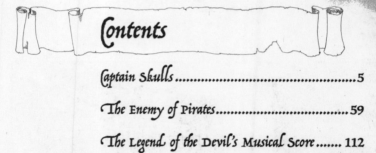

Contents

WANTED

Story & Art by Matsuri Hino